Catchment Hydrological Modelling

Catchment Hydrological Modelling

The Science and Art

Shreedhar Maskey

Elsevier
Radarweg 29, PO Box 211, 1000 AE Amsterdam, Netherlands
The Boulevard, Langford Lane, Kidlington, Oxford OX5 1GB, United Kingdom
50 Hampshire Street, 5th Floor, Cambridge, MA 02139, United States

ISBN: 978-0-12-818337-3

For information on all Elsevier publications
visit our website at https://www.elsevier.com/books-and-journals

Publisher: Candice Janco
Acquisitions Editor: Louisa Munro
Editorial Project Manager: Sara Valentino
Production Project Manager: Sruthi Satheesh
Cover Designer: Victoria Pearson Esser

Typeset by STRAIVE, India

Contents

Acknowledgments

At IHE Delft, I have had the privilege to work with many MSc and PhD students from different countries and regions, most of whom have used catchment hydrological models, one way or the other, in their research projects. Working with them, I learned the ins and outs of many different hydrological models, which gave me the enthusiasm to write this book. I am grateful to all students and researchers who I have worked with.

I would like to thank Louisa Munro, Sara Valentino, and Sruthi Satheesh at Elsevier, who have been very helpful at the different stages of the publishing process from submitting the book proposal to finalizing copyediting. I very much appreciate their understanding and consideration when I had to ask for more time to finalize the chapters and corrections.

I am also thankful to many of my colleagues at IHE; those who knew I was writing the book were always encouraging and supportive. Special thanks to my colleagues Raymond Venneker and Yangxiao Zhou, who kindly and thoroughly reviewed two chapters (Evaporation and Groundwater) of the book.

I started writing the book and completed a chapter when I was on a sabbatical at the Graduate School of Advanced Integrated Studies in Human Survivability of Kyoto University. I am very grateful to Professor Kaoru Takara, who was then the dean of the graduate school, for kindly hosting me there with excellent office facilities. In my meetings with him, I learned a lot from his wealth of knowledge and experience on a broad range of topics. I am also thankful to Professor Vincent Guinot at the University of Montpellier for giving me useful feedback and practical advice on writing the book.

Last but not least, I am extremely grateful to my family (wife and children) and my sisters and brother and other loved ones in Nepal for their love, support, and understanding. In the last two years, I spent most of my holidays and weekends writing the book. I owe my family an apology for spending little time with them during these two years.

CHAPTER

Introduction

1

1.1 What is a catchment hydrological model?

All of us have some idea about what a catchment or hydrological model is, but how we actually define it may vary. In a way, observing a catchment model running in a computer is like observing mathematics at work. When we run a model, we execute a set of mathematical equations. This also means that throughout this book a 'model' refers to a mathematical model, unless it is specifically defined otherwise in the given context. Mathematical equations describe how something works or behaves under certain conditions. So the equations of a catchment hydrological model should describe the behaviour of the hydrological processes and their interactions in the catchment.

The terms 'catchment model' and 'hydrological model' are used as synonyms in this book. Occasionally to emphasize both 'catchment' and 'hydrology', the term 'catchment hydrological model' is also used synonymously. Sometimes models are also distinguished as 'hydrological' and 'hydraulic' models, but this distinction is not intended when the term 'hydrological model' is used in this book. In the literature a hydrological model is also commonly referred to as a 'rainfall–runoff' or 'precipitation–runoff' model, where the emphasis is on the major input and output variables of the model, that is, rainfall (or precipitation) and runoff. Note, however, that the output of a hydrological model is not limited to only runoff, so a catchment or hydrological model can be viewed in a broader sense than a rainfall–runoff model.

To understand how a catchment model works, we can ask a basic question: how are the equations of a catchment model formulated? In the model formulation process, we start with forming our view or concept and assumptions about the functioning of the catchment's hydrological processes. For example, it is easy to conceive that when rain falls on the ground, part of it infiltrates into the soil if the surface is permeable, and part of it may evaporate. This may sound very simple, but it immediately draws us to the next level of process detail or connections between processes. For example, what happens with the infiltrated water? Or, how can we know how much of the rain water evaporates and how much of it infiltrates? And so on. In the literature, this stage of model formulation is referred to as "conceptual model" formation (Blöschl and Sivapalan, 1995; Gupta et al., 2012) or "perceptual model" formation (Beven, 2001). Note that the term "conceptual model" is also used in the classification of models based on the type of methods/equations used to represent the physical processes, which is discussed in the next section.

Catchment Hydrological Modelling. https://doi.org/10.1016/B978-0-12-818337-3.00001-5

The next stage is to find the right equation to describe these processes and connections. Broadly, we can think of three ways to formulate equations to describe the hydrological processes in a catchment. First, there are equations that are based on fundamental physical laws, in particular the conservation laws: conservation of mass, conservation of momentum and conservation of energy. For example, the water balance equation represents the conservation of mass, the evaporation equation can be derived from the energy balance, and the Saint Venant equations of water flow in a channel are a set of two equations describing the conservation of mass and momentum. We will talk about these equations later in the book. While describing catchment processes using mathematical equations, it becomes apparent that not everything that happens in a catchment can be practically described solely using the fundamental physical laws. In the second type of equations, we try to find a relationship that still attempts to capture or describe some aspects of the physical process in a "reasonable" but simplified way (see, e.g. Dingman, 2002). An example of this type of equation is the linear reservoir method. This type of relationship (equation) is also called a conceptual model or method, which has been proven to be a very useful method in catchment modelling (see Chapters 4–6). The third type of equation is derived from experimental/observation data to represent how one variable is related to one or more other variables, so these are empirical equations. A good example of such an equation is the temperature index method for snowmelt (described in Chapter 8), which relates the amount of a day's snowmelt to the temperature of the day exceeding the melting threshold temperature.

The term 'model' is also used to refer to the catchment model as a whole as well as the model of a particular process or component of the catchment model. For example, an evaporation model refers to a mathematical model of the evaporation process, which is a component in a catchment model. The process of formulating a mathematical model discussed here is the same for a component model (such as evaporation model, unsaturated zone model) and a catchment model as a whole.

1.2 Types of models

Whether we need to choose a hydrological model or we are already using one for modelling a catchment, we probably want to know what type of model we are going to choose or we are using. The problem is we may not find a clear answer to this question. The reason is because hydrological models can be classified in different ways, and sometimes the distinction between model types is vague or inexplicit. In Section 1.1 I said that the 'model' referred to in this book is a mathematical model. So, one classification of a model is also a 'mathematical model' or a 'physical model'. In the mathematical model, we represent the physical catchment in the computer through a set of input data, and its behaviour or physical processes through mathematical equations. In the computer, the catchment is basically numbers and equations. Sometimes, mathematical models are also called computer models. The physical model, on the other hand, is a real physical object we build in the laboratory as a scaled-down version of the real catchment. Physical models are usually built to study some specific phenomena that may be difficult to accurately represent in a mathematical model. Three different classifications of mathematical models that are commonly used in the hydrological modelling literature are discussed here. These are (1) physically based, conceptual and empirical models; (2) lumped, distributed, and semidistributed models; and (3) Continuous-time simulation and event-based simulation models.

These classifications are also helpful to better understand a hydrological model. The basis for the first two classifications (above) is related to the physical process representation in the model—how and with what details and simplifications. The third classification is related to the time scale in which the models are expected to run. More information about model classifications can be found in Singh (1995) and Dingman (2002).

1.2.1 Physically based, conceptual and empirical models

The classification of hydrological models as *physically based*, *conceptual* and *empirical* is based on how the hydrological processes that exist in the real catchment are mathematically represented in the model. This is about formulating the type of equations (also discussed in Section 1.1) to represent the relevant processes in the catchment, e.g. the evaporation process, surface and subsurface flow processes.

In the *physically based* model, the pertinent processes are represented using partial differential equations of the conservation of mass, energy and momentum (Freeze and Harlan, 1969; Abbott et al., 1986). There are two aspects here: the conservation laws and the partial differential equation. The conservation of mass (water balance) is arguably the single most important concept in catchment models. More details about the water balance are discussed in Section 1.6. For a hydrological model the energy balance refers to the balance between the radiative energy (incoming and outgoing radiation) and the latent heat energy (i.e. the energy required to transfer liquid water to vapour or ice to liquid water) and sensible heat energy transfer. The energy balance is relevant to two important processes in hydrological models: evaporation and snow/ice melt. We will discuss the energy balance equation in more detail in Chapters 3 and 8. The conservation of momentum represents the balance of forces, which is applied in the derivation of water flow equations, for example, surface flow and river flow equations (Chapters 5 and 7). The famous Darcy's law applied in the groundwater flow is also related to the momentum principle (see, e.g. Bear, 1972).

The "physically-based, digitally-simulated" concept for the hydrological model was first proposed in a seminal paper by Freeze and Harlan (1969). However, due to its apparent complexities, despite the exponential advances in the computational (computer) capacity and techniques since then, the development and application of comprehensive physically based models is still limited. The European Hydrological System [Systeme Hydrologique Europeen (SHE); Abbott et al., 1986], now called MIKE SHE (DHI, 2017), is probably the most well-known physically based hydrological model (Montanari and Koutsoyiannis, 2012). However, its application is also limited compared to some other comprehensive conceptual models that are open access.

In the *conceptual model*, the approach is to include as much as possible the processes that take place in the catchment as distinguishable components and the connections between them, but the processes are represented using more simpler algebraic equations as opposed to mostly partial differential equations in the physically based models. The details with which the relevant factors of the processes are encapsulated also vary substantially among different models. To give one example, we said earlier what happens with rainfall when it reaches the land surface. Part of it evaporates, part of it may remain in surface depressions (for some time), and part of it infiltrates through the soil. If the rainfall continues, some or all of it may become the surface runoff. In a detailed conceptual model, all of these processes and their interlinkages are considered as in a detailed physically based model, but with a different approach or equations to represent them. The soil infiltration process, which is one of the most difficult processes to accurately simulate, is a good example. A physically based model would use the partial differential Richards equation (see Chapter 4) and attempt to represent the physics of the water movement in the soil. In a conceptual model, it is represented in a more simpler way, usually assuming one or more soil layers with each layer's maximum water content and maximum infiltration rate, for example. Then the actual amount of infiltration would be determined based on how much water is available and what is the potential of a layer for infiltration. The soil water (moisture) accounting model (Crawford and Linsley, 1966; Burnash et al., 1973) is a good example of a conceptual catchment model. Examples of conceptual models with different levels of detail are found in Fowler et al. (2016).

In the *empirical model*, the relationship between the input and the output is derived usually with observation data with little or no attempt to apply the knowledge of the physical processes involved. The unit hydrograph model to transfer the excess rainfall to river discharge (see Chapter 5) is an example of an empirical method commonly used with catchment models. Some of the empirical methods used in hydrology are also referred to as "system-theoretical" models (Lettermann, 1991). Linear and nonlinear regressions are commonly used with empirical models. Models that use more sophisticated techniques for regression, e.g. artificial neural network (French et al., 1992), fuzzy regression (Bardossy et al., 1990; Bárdossy and Disse, 1993) and support vector machine (see, e.g. He et al., 2014), are more commonly called *data-driven* models, but technically they are a family of empirical models.

The model distinctions physically based, conceptual and empirical are generally suited to describe a model of a particular process or component. When it comes to the full catchment model, many models use methods/equations some of which may qualify for a physically based, some conceptual and some empirical. In such a case, it is more about whether a particular model has characteristics with more resemblance to a particular type than about whether it is physically based, conceptual, or empirical. If a model has most of the process details represented but some are fully or loosely physically based and some are conceptual/empirical, it may be more appropriately called a *process based* model as opposed to physically based or conceptual (see, e.g. Montanari and Koutsoyiannis, 2012). However, note that *physically based* and *process based* are also used synonymously. More comprehensive discussions on process based/physically based are found in Montanari and Koutsoyiannis (2012) and Clark et al. (2015).

1.2.2 Lumped, distributed and semidistributed models

Hydrological models are classified as *lumped*, *distributed* or *semidistributed* based on the details with which the spatial variation of the processes and the catchment's physical properties are represented in the model. A fully *lumped model* is very simple: it takes one value (spatially averaged) for each input variable and model parameter for the entire catchment. The input may vary in the time domain. For example, precipitation usually varies over the catchment, but a lumped model takes precipitation for a given time step same everywhere in the catchment as one areal average value. Similarly, the catchment may have different land cover types (e.g. forest land, crop land and urban area), with different hydrological properties, for example, related to infiltration and surface runoff. In a lumped model, the different values of the same property are lumped into a single value.

In a *distributed model*, spatial variation of the model inputs and parameters can be represented. Usually the distributed model is constructed with a grid (or mesh) structure (e.g. a rectangular grid) such that by varying the grid size the distributed inputs can be specified with an intended level of detail. But, of course, the data with the required details need to be available too. Note, however, that the grid size can vary, and so if a large grid size is used and the values are averaged for each grid cell, the distributed (grid-based) structure does not necessarily mean the process representation is distributed. On the other hand, a very small grid size means too many grid cells and too many computational units, which then requires large computational time and may adversely affect calibration efficiency (also discussed in Chapter 9). To give an example, in one study, a distributed hydrological model was developed for the head water catchment of the Yellow River (also called the upper Yellow River) using a 5 × 5 km horizontal grid resolution for the modelled area of about 125,000 km^2 (Maskey et al., 2007; see Hu et al., 2011 for a more detailed description of the catchment). This means 5000 grid cells, which is equivalent to using 5000 lumped models of 25 km^2 area each, plus the additional computations to route water from cell to cell. This was not the finest resolution model for the catchment; later Hu (2014) set up another distributed model with 1 × 1 km resolution, which means 25 times more grid cells.

In the *semidistributed* model structure, the catchment is usually divided into a number of subcatchments—using the drainage area boundary as its spatial unit instead of square/rectangular grids. This is hydrologically logical and usually a convenient structure for a catchment level analysis. By varying the size and number of subcatchments, the semidistributed model structure also allows to represent the variation of inputs and parameters at the desired level of details. Probably, a large majority of hydrological model applications are based on semidistributed conceptual (or process based) models. To name a few, applications of different semidistributed models can be found in Biftu and Gan (2001), Pechlivanidis et al. (2010), Shrestha et al. (2013), and Wortmann et al. (2019). Moreover, some models also divide sub-catchments into smaller areas, generally called hydrological response units (HRUs), using a unique combination of catchment physical characteristics, such as land use, soil type and topographic slope (see, e.g. Sirisena et al., 2018). In this way, models with a semidistributed structure can be made more distributed to represent the spatial variability.

1.2.3 Continuous-time simulation and event-based simulation models

Hydrological models are also distinguished as *continuous-time simulation* and *event-based* on the basis of the intended time scale to run the model. The *event-based* model refers to the models that are intended to run for isolated rainfall events. These models are aimed at simulating the runoff response of a catchment to a particular rainfall event and have applications in, e.g. flood forecasting and storm discharge estimation for urban drainage design. The *continuous-time simulation* model refers to the models that are intended to run for the entire time period of simulation without differentiating the rainfall events, and as such these models should be capable of simulating sequences of wet and dry or high and low flow periods continuously.

From the model building point of view, there are two main differences between the event-based and continuous-time simulation models. The first difference is about the type of the methods to choose for a particular process or model component. Some methods are intended and effective only for event-based simulation, while some are intended for continuous simulation and may be less effective for event-based simulation. A good example is the surface runoff volume estimation by the Soil Conservation Service Curve Number (SCS CN) and the Soil Moisture Accounting (SMA) methods (see Chapter 4). The SCS CN method is primarily developed for flood discharge estimation from a rainfall event, although with the use of antecedent moisture conditions and combined with the continuous groundwater (baseflow) method it has generally been used for continuous simulation successfully. See, for example, the Soil and Water Assessment Tool (SWAT; Neitsch et al., 2011), which is a comprehensive catchment hydrological model. The SMA method is primarily intended for continuous-time simulation and may face challenges to effectively simulate individual events.

Another difference is related to the model initialization and calibration strategy. In the continuous-time simulation, a standard practice is to run the model for a certain period of time, also called a warm-up period, before the intended simulation period. For the long-term simulation, the warm-up period is usually a year or longer. In the event-based simulation, the warm-up period strategy may not be appropriate, and so the model's initial state may need to be determined in a different way, e.g. using observation data.

1.3 Basin, catchment, watershed—Are they all the same?

We often find the terms 'basin', 'catchment', and 'watershed' used interchangeably when referring to a river or river system and its drainage area. Dictionary definitions of these terms may vary slightly, but when coupled with a river name (e.g. Nile basin, Nile catchment or Nile watershed), or with another noun (e.g. basin area, catchment area or watershed area), it is primarily a matter of author's preference to use one or the other (see Box 1.1).

However, when two of the terms are used in the same context, some distinctions are made. For example, when the terms basin and catchment are used to distinguish one subarea (or one type of the catchment area) from a larger catchment area, authors seem to use 'basin' for the larger area and 'catchment' for the subarea, e.g. in "...the Headwater *Catchment* of the Yellow River *Basin*..." (Wang et al., 2019) and in "... for *catchments* in two Indian river *basins*" (Chavan and Srinivas, 2015). Similarly, when the terms basin and watershed are used together, 'basin' for the larger area and 'watershed' for the subarea, or basin with the river name and watershed with area characteristics are commonly used, e.g. in "...a Semi-Arid, Agricultural *Watershed*: Yakima River *Basin*..." (Grieger and Harrison, 2021) and in "...a boreal forest *watershed*: White Gull Creek *basin*..." (Nijssen and Lettenmaier, 2002). When the terms watershed and catchment are used together, there is no single pattern. Some authors use 'watershed' for the larger area and catchment for the subarea,

Box 1.1 How often are 'basin', 'catchment' and 'watershed' used with river names?

Out of curiosity, I searched on Google several river names combined with 'basin', 'catchment' and 'watershed' (search date: 31 October 2021). The river names searched were Amazon, Mississippi and Colorado from South and North America; Nile and Niger from Africa; Danube and Loire from Europe; and Ganges, Bagmati, Yellow River, Ishikari, Moon River and Murray-Darling from Asia and Australia. With all of these river names, the number of search results (hits) was overwhelmingly high for 'basin'. For example, the search results for 'Nile basin' were 2.98 million compared to just 14,300 for 'Nile catchment'. With Amazon, it was 4.62 million search results for 'Amazon basin' compared to 33,000 for 'Amazon watershed'. The second highest number of search results was for either 'catchment' or 'watershed' depending on the river name. The second highest search results found were for 'watershed' for the rivers from North and South America (Amazon, Mississippi and Colorado) and for 'catchment' for the rivers from Africa and Europe (Nile, Niger, Danube and Loire). With Asian rivers the second highest hits were mixed: for 'watershed' for Ganges, Bagmati, Yellow and Moon rivers and for 'catchment' for Ishikari and Murray Darling rivers.

In scientific journal publications, I searched the titles of all papers published in the *Journal of Hydrology* (JoH), *Hydrology and Earth System Sciences* (HESS) and *Water Resources Research* (WRR). In JoH and WRR, the highest search results were for 'basin', but the differences were relatively small: in WRR, 550, 536 and 512 and in JoH, 1205, 1116, and 634 for 'basin', 'catchment' and 'watershed', respectively. In HESS, the highest results were for 'catchment' (601) compared to 'basin' (495) and 'watershed' (163).

e.g. in "...nested *catchments* in a temperate forested *watershed*" (James and Roulet, 2006) and in "...urban-rural *catchments* in the Chesapeake Bay *watershed*" (Shields et al., 2008), and also interchangeably, e.g. in Jeelani et al. (2010). In Rose (2006), the terms 'drainage basin' and 'catchment' are used interchangeably, but 'watershed' is used for a subunit within a catchment, so a watershed instead of a subcatchment.

In this book 'catchment' and 'subcatchment' are used more or less consistently, also as a matter of preference. Occasionally when used with a river name, 'basin' is also used, e.g. Indus basin.

1.4 Purpose of a catchment model

Catchment hydrological models can be used in at least four different ways. (Note that this is just one of many ways a model's uses can be classified.)

1. To reproduce the catchment's response in terms of water fluxes and storage changes to the historical meteorological data.
2. To predict the catchment's response in terms of the fluxes and storage changes in the long-term future. The long-term future typically refers to the time scale of years to decades.
3. To forecast the fluxes and storage changes in the short-term future. The short-term future usually means the time scale of days. Forecasts in the range of weeks or months may be called medium-term.
4. As a scientific exercise, to test the model or model hypothesis, compare models, test modelling-related techniques, such as calibration, and uncertainty prediction.

With the historical period meteorological data, hydrological models can be used broadly for two objectives. First, to analyse or investigate the fluxes or storage changes that were not measured in the catchment, such as river discharge, evaporation and groundwater storage. In many water resources applications, for example, water resources planning and hydropower design, long-term discharge data are required. In many river catchments, discharge data are not measured (ungauged basins) or only sparsely measured. Given that the global precipitation data are readily available these days (see Chapter 2), hydrological models can be used to derive catchment runoff in ungauged or poorly gauged catchments (Sirisena et al., 2018). Also related to this is the use of hydrological models in regionalization studies (Masih et al., 2010) in ungauged or partially gauged catchments. Second, to assess how the catchment would respond to interventions (in the form of land use change or reservoirs, diversions, etc.) under the given climatic conditions.

The prediction of the catchment's response in the long-term future is similar to assessing the impacts of changes in the historical period but under the future climate. Over the past two decades hydrological models have been extensively used in climate change impact assessments using climate projections from global circulation models. Understandably applications for the future periods are associated with more challenges and uncertainties than applications with the historical data (see also Chapter 9).

The short-term and medium-term forecasts are typically carried out for flow forecasting for various water uses (e.g. hydropower, irrigation and drinking water supply) and for flood and drought forecasting. Forecasting is one of the most common applications of hydrological models.

1.5 What makes a catchment model different from a river model?

A catchment hydrological model is built representing all the drainage area up to a selected outlet point at a river or stream in the catchment. Rivers/streams are an integral part of a catchment. If a hydrological model is to simulate water flow in a catchment, one way and another, it has to simulate the flow in the rivers. So then why we need a separate model only for the river, or what makes a catchment model different from a river model? We can also ask a reversed question, from the river model's perspective: if rivers are an integral part of a catchment, then how can we model a river without modelling its catchment?

In a catchment model, the output of interest is often at the outlet point of a catchment or subcatchment. In most catchment models that are semidistributed or distributed, river flow routing is represented as one of the model components. In this case the only difference is in the methods used to model the rivers in the catchment model and in the separate river model.

In a separate river model, the flow in the river are represented using the conservation of mass and momentum. The methods that represent the conservation of momentum can be of different level of details. For example, the Saint Vanent equations (which are presented in Chapter 7) represent the fully dynamic flow routing often used in a one-dimensional river model. When the conservation of mass and momentum are used, the channel geometry (e.g. cross section and slope) as well as the channel bed resistance to flow are also included. This allows to compute flow velocity, discharge and water depth in the channel. Simplified methods include the kinematic wave and Muskingum-Cunge routing, which is a variant of the kinematic wave method.

In catchment models, river flow may be modelled based on only the conservation of volume, e.g. the Muskingum method (Chapter 7), which allows to compute the change in discharge along the river reach without computing the detailed flow characteristics (e.g. water depth or velocity in the channel). In a catchment model, even if the rivers are modelled with kinematic wave or Muskingum-Cunge methods, usually a lot simplified (approximated) channel geometry is specified and consequently water depths or velocities are not considered as primary output variables. If the rivers are not represented as a separate element in the catchment model, it may use indirect techniques to take into account the effects of the flow in the river channel on the estimated flow at the outlet point.

To model only the river of a catchment, the river needs to be separated from the catchment. But by definition the rain water from the catchment drains into the river system in the catchment. So, the river must be supplied with water in the same way the catchment would drain water into it. In a river model this is done through the upstream boundary condition: either a point boundary if the water input is at a point, or a distributed boundary if the input is along certain length of the river. The upstream boundary is typically a hydrograph (discharge versus time).

The third approach of modelling a catchment–river system is to build catchment and river models separately and then couple them dynamically during run-time. In such coupled modelling, references are to be made to specify which points of the rivers (in the river model) are connected to which grid-cell or subcatchment outlets in the catchment model. At each time step, the running sequence is first to run the catchment model (to generate necessary inflow or upstream boundary input for the river model) and then run the river model. However, there is one problem to run the catchment and river models in a sequence, because it is not always the case that the catchment drains water into the river but also the river may supply rechange to the aquifer system in the catchment. One "standard" trick usually used to handle this problem is to estimate the aquifer—river flow exchange based on the river water level from the previous time step, which is similar to using an explicit method in numerical solutions (see e.g. Chapter 5).

1.6 Components of a catchment model and catchment water balance

In a hydrological model, the processes of water movement in a catchment are simulated starting from precipitation that falls on the catchment. The precipitation formation process itself is beyond the scope of a catchment hydrological model. The major processes and model components typically considered in a hydrological model include canopy interception, evaporation and transpiration, unsaturated zone flow, saturated zone (or groundwater) flow, surface flow routing, and river flow routing, and additionally snow accumulation and snow/ice melt for catchments with snow precipitation. These components and methods to model them are presented in Chapters 3–8.

This book emphases on understanding how each of these components is dependent on other components for input and output, because only then we can comprehend the details that go into building a catchment model. Understanding the relationships between model components (representing hydrological processes) is also necessary to correctly interpret model results and to effectively diagnose problems usually encountered in model calibration (see Chapter 9). In the chapter where each model component is described, its connections with other components for input and output are also discussed. The overall connection of all the components in a full catchment model is presented in Chapter 9, which comes after each individual component is presented, because by then it should be more convenient to see the role of each component in the performance of the catchment model. The overall connection presented in Chapter 9 represents, in a way, the hydrological cycle but with more detailed indication of how the individual components are dependant on each other.

As said earlier, the water balance is about applying the principle of conservation of mass of water coming in and going out from the catchment. The water balance can be applied for different components of the catchment model as well as for the entire catchment. For example, in Chapter 6, the water balance of groundwater aquifer is presented and in Chapter 8, mass balance of snow accumulation/melt is presented. In short, water balance equation is the Change in Storage (volume per time) = Inflow – Outflow. The simplest form of the catchment water balance equation can be written as:

Change in storage over time (dS/dt) =
+ Precipitation (P)
– Evaporation (E)
– Streamflow (or runoff) measured at the outlet of the catchment (Q).

That is

$$\frac{dS}{dt} = P - E - Q \tag{1.1}$$

where S is the storage volume [L^3], t is time [T], and P, E and Q are in volume per time [$L^3 T^{-1}$]. The change in storage dS/dt refers to the change in all types of storage in the catchment over the time period. These include the groundwater storage, soil moisture storage, interception storage, snow/ice storage, and lake/reservoir storage.

In practice, quantities P and E are commonly expressed as depth of water per time [LT^{-1}], usually in mm day^{-1}. So, Eq. (1.1) can be expressed in two alternative ways. Either with only P and E in depth per unit time (Eq. 1.2):

$$\frac{dS}{dt} = A_{cat}(P - E) - Q \tag{1.2}$$

where A_{cat} is the catchment area in [L^2], P and E are in depth per unit time [LT^{-1}], Q is in volume per unit time [$L^3 T^{-1}$], and S and t are in the same unit as in Eq. (1.1).

Or with all three (P, E and Q) in depth pet unit time (Eq. 1.3):

$$\frac{1}{A_{cat}}\frac{dS}{dt} = P - E - Q \tag{1.3}$$

where A_{cat} is the catchment area in [L^2] (same as in Eq. 1.2), P, E, and Q are all in depth per unit time [LT^{-1}], and S and t are in the same unit as in Eqs. (1.1) and (1.2).

The water balance Eqs. (1.1) to (1.3) assume that the catchment has only one inflow source as precipitation, and two outflows: one as evaporation and one as river discharge at the outlet of the catchment. Because the catchment area boundary is based on the surface topography, the groundwater aquifer may not be fully contained in the catchment topographic boundary. In such a case, the groundwater may have additional inflow and outflow sources (see Chapter 6).

Another situation is that the catchment may have a groundwater aquifer (usually a deep aquifer) which is not connected to the river system in the catchment where the discharge (Q) is measured, but the aquifer may receive recharge from the catchment area. If such aquifer recharge exists in the catchment but is not accounted for in the model can lead to erroneous water balance. Note that some hydrological models, e.g. SWAT and HEC-HMS (USACE, 2000), allow deep aquifer recharge and consider it as a loss term (additional outflow) in the catchment water balance.

1.7 The science and art of catchment modelling

In the context of hydrological modelling, I came across the idea of science and art when I read an opinion paper by Hubert Savenije, Professor of Hydrology, in 2009, long before I dwelled on writing this book. The title of his paper reads "The art of hydrology" (Savenije, 2009). I know Hubert Savenije personally—a very friendly person, whose lectures are fascinating and arguments persuasive. The science-and-art idea stayed with me ever since, but it was not until 2018 when I was writing the proposal for this book and also reading the biography of Leonardo Da Vinci (Isaacson, 2017) that I decided to use this title. Da Vinci was best known to most people for his paintings, which were unquestionably superb, but he was also an engineer and a scientist (Isaacson, 2017). He is an exceptional example of a human being who had the highest level of imagination and creativity of an artist and the deepest level of curiosity and passion of a scientist.

What is science and what is art anyway? In art the emphasis is on "imagination" and "creativity", and in science it is the systematic "observation" and "experiment". The Oxford Dictionary (2010) defines art as "the expression and application of human creativity and imagination" and science as "... the systematic study of ... the physical and natural world through observation and experiment."

However, is it fair to separate science and art like this or to differentiate a scientist's and an artist's qualities like this? I guess no one will disagree when I say that a good scientist also uses imagination and creativity. I do not think Einstein would have been the Einstein or Newton the Newton without the possession of the highest level of imagination. It is unfair to separate science and art all the time. Science needs art to succeed at the highest level.

I was on a sabbatical from June to September 2019 at the Graduate School of Advanced Integrated Studies in Human Survivability (SHISHU-KAN) of Kyoto University when I seriously started writing this book. When I was discussing the book and its title with Prof. Kaoru Takara, who was then the dean of the Graduate School, he explained to me about what we traditionally know as STEM (Science, Technology, Engineering and Mathematics) education and the emerging STEAM (Science, Technology, Engineering, Arts and Mathematics) education (intruducing arts into STEM) (see, e.g. Oxford University Press, n.d.)[42].

Savenije (2009)'s argument about the role of the "art" in hydrology and hydrological modelling lies in its "ability to reconstruct the architecture of a largely unknown system from a few observable signatures." Imaginative power and creative thinking are necessary to conceptualize the complex hydrological processes—flow of water in highly heterogeneous media and exchange of mass and energy fluxes—at the catchment scale and effectively relate them with the scientific theories developed from the observable and measurable scale. In a commentary paper "The art of catchment modeling…", Barnes (1995) also argued about the role of arts in hydrological modelling. His main point is that hydrological modelling is a subjective exercise in many ways (e.g. for the selection of the type of models and model parameterization) and that to develop a successful model requires "subjective judgement based on experience."

More recently, Ramos and Arnal (2018) wrote a compelling blog post in trying to find "art" in hydrological forecasting. A lot is said in the post, but I would add or stress on one thing. That in flood forecasting, the role of an artist's mind and skill are crucial, not least in creating a visualization to communicate the value of the forecasts however uncertain they may be.

A lot has been said in Savenije (2009) and Barnes (1995) on why hydrological modelling is "art as much as it is science". My intention in this book is not to demonstrate per se the presence of art or science in hydrological modelling, but to argue that imagination and creativity, which are more the qualities of an artist, can benefit hydrological modelling in many ways. To name a few: for choosing an appropriate balance between conceptual and process based model components in view of the purpose of the model and data availability, effective model calibration combining manual and automatic calibration (see Chapter 9), and interpreting the potentially uncertain model result in such a way that it is understandable and useful to the intended users.

To conclude, I quote what Walter Isaacson has said in Da Vinci's biography "[Leonardo] knew that art was a science and that science was an art."

References

Abbott, M.B., Bathurst, J.C., Cunge, J.A., O'Connell, P.E., Rasmussen, J., 1986. An introduction to the European hydrological system – Systeme Hydrologique Europeen, "SHE", 2: structure of a physically-based, distributed modelling system. J. Hydrol. 87, 61–77.

Bárdossy, A., Disse, M., 1993. Fuzzy rule-based models for infiltration. Water Resour. Res. 29 (2), 373–382. https://doi.org/10.1029/92WR02330.

Bardossy, A., Bogardi, I., Duckstein, L., 1990. Fuzzy regression in hydrology. Water Resour. Res. 26 (7), 1497–1508.

Barnes, C.J., 1995. The art of catchment modeling: what is a good model? Environ. Int. 21 (5), 747–751.

Bear, J., 1972. Dynamics of Fluids in Porous Media. Dover Publications, Inc., New York.

Beven, K.J., 2001. Rainfall-Runoff Modelling: The Primer. Wiley, Chichester.

Biftu, G.F., Gan, T.Y., 2001. Semi-distributed, physically based, hydrologic modeling of the Paddle River basin, Alberta, using remotely sensed data. J. Hydrol. 244, 137–156. https://doi.org/10.1016/S0022-1694(01)00333-X.

Blöschl, G., Sivapalan, M., 1995. Scale issues in hydrological modelling: a review. Hydrol. Process. 9 (3–4), 251–290. https://doi.org/10.1002/hyp.3360090305.

Burnash, R.J.C., Ferral, R.L., McGuire, R.A., 1973. A Generalized Streamflow Simulation System—Conceptual Modeling for Digital Computers. U.S. Department of Commerce, National Weather Service and State of California, Department of Water Resources.

Chavan, S.R., Srinivas, V.V., 2015. Effect of DEM source on equivalent Horton–Strahler ratio based GIUH ratio based GIUH for catchments in two Indian river basins. J. Hydrol. 528, 463–489. https://doi.org/10.1016/j.jhydrol.2015.06.049.

Clark, M.P., et al., 2015. A unified approach for process-based hydrologic modeling: 1. Modeling concept. Water Resour. Res. 51, 2498 2514. https://doi.org/10.1002/2015WR017198.

Crawford, N.H., Linsley, R.K., 1966. Digital simulation in hydrology: Stanford Watershed Model IV. Technical report no. 39, Department of Civil Engineering, Stanford University, Stanford, California.

DHI (2017). MIKE SHE Volume 2: Reference Guide. DHI, Denmark. https://manuals.mikepoweredbydhi.help/2017/Water_Resources/MIKE_SHE_Printed_V2.pdf; Accessed on 31 July, 2021.

Dingman, S.L., 2002. Physical Hydrology, second ed. Prentice Hall, New Jersey.

Fowler, K.J.A., Peel, M.C., Western, A.W., Zhang, L., Peterson, T.J., 2016. Simulating runoff under changing climatic conditions: revisiting an apparent deficiency of conceptual rainfall-runoff models. Water Resour. Res. 52, 1820–1846. https://doi.org/10.1002/2015WR01806.

Freeze, R.A., Harlan, R.L., 1969. Blueprint for a physically-based, digitally simulated hydrologic response model. J. Hydrol. 9, 237–258.

French, M.N., Krajewski, W.F., Cuykendall, R.R., 1992. Rainfall forecasting in space and time using a neural network. J. Hydrol. 137, 1–31.

Grieger, S.R., Harrison, J.A., 2021. Long-term disconnect between nutrient inputs and riverine exports in a semi-arid, agricultural watershed: Yakima River Basin 1945–2012. J. Geophys. Res. Biogeo. 126. https://doi.org/10.1029/2020JG006072, e2020JG006072.

Gupta, H.V., Clark, M.P., Vrugt, J.A., Abramowitz, G., Ye, M., 2012. Towards a comprehensive assessment of model structural adequacy. Water Resour. Res. 48, W08301. https://doi.org/10.1029/2011WR011044.

He, Z., Wen, X., Liu, H., Du, J., 2014. A comparative study of artificial neural network, adaptive neuro fuzzy inference system and support vector machine for forecasting river flow in the semiarid mountain region. J. Hydrol. 509, 379–386. https://doi.org/10.1016/j.jhydrol.2013.11.054.

Hu, Y., 2014. Water Tower of the Yellow River in a Changing Climate: Toward an Integrated Assessment. PhD Thesis, CRC Press/Balkema, Leiden.

Hu, Y., Maskey, S., Uhlenbrook, S., Zhao, H., 2011. Streamflow trends and climate linkages in the source region of the Yellow River, China. Hydrol. Process. 25, 3399–3411.

Isaacson, W., 2017. Leonardo Da Vinci—The Biography. Simon & Schuster, London.

James, A.L., Roulet, N.T., 2006. Investigating the applicability of end-member mixing analysis (EMMA) acrossscale: a study of eight small, nested catchments in a temperate forested watershed. Water Resour. Res. 42, W08434. https://doi.org/10.1029/2005WR004419.

Jeelani, G., Bhat, N.A., Shivanna, K., 2010. Use of δ18O tracer to identify stream and spring origins of a mountainous catchment: a case study from Liddar watershed, Western Himalaya, India. J. Hydrol. 393, 257–264. https://doi.org/10.1016/j.jhydrol.2010.08.021.

Lettermann, A., 1991. Systerm-Theoretic Modelling in Surface Water Hydrology. Springer-Verlag, Berlin.

Masih, I., Uhlenbrook, S., Maskey, S., Ahmad, M.D., 2010. Regionalization of a conceptual rainfall-runoff model based on similarity of the flow duration curve: a case study from the semi-arid Karkheh basin. Iran. J. Hydrology 391, 188–201.

Maskey, S., Venneker, R., Uhlenbrook, U., 2007. Large-scale flow modelling application in the Yellow River Basin using satellite derived input data. In: The 2nd International Conference on GIS/RS in Hydrology, Water Resources and Environment (ICGRHWE'07) and the 2nd International Symposium on Flood Forecasting and Management with GIS and Remote Sensing (FM2S'07), 7–13 September 2007, Guangzhou and Three Gorges, China.

Montanari, A., Koutsoyiannis, D., 2012. A blueprint for process-based modeling of uncertain hydrological systems. Water Resour. Res. 48, W09555. https://doi.org/10.1029/2011WR011412.

Neitsch, S.L., Arnold, J.G., Kiniry, J.R., Williams, J.R., 2011. Soil and Water Assessment Tool Theoretical Documentation Version 2009. Texas Water Resources Institute. Available electronically from https://hdl.handle.net/1969.1/128050.

Nijssen, B., Lettenmaier, D.P., 2002. Water balance dynamics of a boreal forest watershed: white Gull Creek basin, 1994–1996. Water Resour. Res. 38 (11), 1255. https://doi.org/10.1029/2001WR000699.

Oxford Dictionary, 2010. Oxford Dictionary of English, third ed. Oxford University Press.

Oxford University Press n.d.. STEAM. https://www.oup.com.au/secondary/STEAM#levels, last accessed 6-November-2021.

Pechlivanidis, I.G., McIntyre, N.R., Wheater, H.S., 2010. Calibration of the semi-distributed PDM rainfall–runoff model in the Upper Lee catchment, UK. J. Hydrol. 386, 198–209. https://doi.org/10.1016/j.jhydrol.2010.03.022.

Ramos, M.-H., Arnal, L., 2018. Can hydrological forecasting contribute to science & art projects? https://hepex.inrae.fr/science-and-art/.

Rose, C.W., 2006. An Introduction to the Environmental Physics of Soil, Water and Watersheds. Cambridge University Press.

Savenije, H.H.G., 2009. HESS opinions "the art of hydrology". Hydrol. Earth Syst. Sci. 13, 157–161. www.hydrol-earth-syst-sci.net/13/157/2009/.

Shields, C.A., Band, L.E., Law, N., Groffman, P.M., Kaushal, S.S., Savvas, K., Fisher, G.T., Belt, K.T., 2008. Streamflow distribution of non–point source nitrogen export from urban-rural catchments in the Chesapeake Bay watershed. Water Resour. Res. 44, W09416. https://doi.org/10.1029/2007WR006360.

Shrestha, B., Babel, M.S., Maskey, S., van Griensven, A., Uhlenbrook, S., Green, A., Akkharath, I., 2013. Impact of climate change on sediment yield in the Mekong River Basin: a case study of the Nam Ou Basin, Lao PD. Hydrol. Earth Syst. Sci. 17, 1–20.

Singh, V.P., 1995. Computer Models of Watershed Hydrology. Water Resources Publications.

Sirisena, T.A.J.G., Maskey, S., Ranasinghe, R., Babel, M.S., 2018. Effects of different precipitation inputs on streamflow simulation in the Irrawaddy River Basin, Myanmar. J. Hydrol.: Region. Stud. https://doi.org/10.1016/j.ejrh.2018.10.005.

USACE, 2000. Hydrologic modelling system HEC-HMS, technical reference manual. In: Feldman, A.D. (Ed.), US Army Corps of Engineers. Hydrologic Engineering Centre, Davis, CA.

Wang, W., Dong, Z., Lall, U., Dong, N., Yang, M., 2019. Monthly streamflow simulation for the headwater catchment of the Yellow River basin with a hybrid statistical-dynamical model. Water Resour. Res. 55, 7606–7621. https://doi.org/10.1029/2019WR025103.
Wortmann, M., Bolch, T., Su, B., Krysanova, V., 2019. An efficient representation of glacier dynamics in a semi-distributed hydrological model to bridge glacier and river catchment scales. J. Hydrol. 573, 136–152. https://doi.org/10.1016/j.jhydrol.2019.03.006.

CHAPTER

Data requirements for a catchment model

2

2.1 Data requirements

Building a catchment hydrological model (mathematical model) is like constructing the catchment in a computer (digitally) in such a way that it behaves or responds to any drivers of change to an acceptable level of similarity as the catchment in the nature does. This is only possible with a good set of data to start with. But what are these data actually? What global data products are available for catchment modelling? What about the quality of these data? How do we use them to build and run the model? What do we know about locally (in situ) measured verse global data products? These are the type of questions discussed in this chapter.

Broadly data requirements for a catchment model can be divided into three categories. These are (1) Time-invariable (or time-invariant) data for model construction (which are generally considered constant over the model simulation period), (2) Time-variable (or time-variant or dynamic) data (which vary during model simulation), and (3) Data for the model calibration, which are also time-variable data.

2.1.1 Time-invariable data for model construction

The time-invariable data are used for model construction and are generally assumed to remain constant during the period of model simulation. Basically these data represent the catchment digitally as a physical object into the model. These are primarily the catchment surface topography usually represented through a DEM (digital elevation model), the soil types and their hydrologically related properties (e.g., texture, depth, porosity, hydraulic conductivity, etc.), land use (land cover) types (e.g., forests of different types, grassland, urban area, agricultural land, etc.), and in case of a model with physically based groundwater component, detailed subsurface information to represent the aquifer system. In a conceptual groundwater model, which is often used with a catchment model, the actual size/shape of the aquifers are not used. The aquifer related data requirements are not discussed here.

Note that depending on the period of time considered, these data may not be always time-invariable. The DEM and soil data and their properties are almost always kept constant in hydrological simulations, but not for all land use/land cover types. Some land cover types and their properties change seasonally, e.g., forests and agricultural lands. So, land use/land covers are usually kept constant in the model,

Catchment Hydrological Modelling. https://doi.org/10.1016/B978-0-12-818337-3.00008-8

but their parameters may vary with seasonality. Usually the seasonality is kept constant year to year. This means that unless the model is specifically for the assessment of land use change impacts, the land use types are constant year to year, but some land use parameters may have intra-annual variations. This is a fair assumption only if any changes in land use that may be expected in the catchment are insignificant for hydrological simulation. There may be situations where the land use change may be substantial and systematic, e.g., large-scale deforestation for agriculture, settlements or other infrastructures. In such a case, either the model should have the possibility to take time-variant land use data in the simulation, or if the change is not gradual but stepwise, the model should be run separately for different periods with the land use data of the respective period.

2.1.2 Time-variable data

The time-variable data refer to the data that typically vary over the simulation period. For a catchment model the time-variable data are primarily the meteorological data, which include precipitation, temperature, net radiation, relative humidity and windspeed. These are also called meteorological forcing data for hydrological model simulation. But not all hydrological models require all of these data. Often the meteorological data are provided on the daily time step. For temperature, usually daily average as well as daily minimum and daily maximum are generally provided. But for some applications meteorological data on a subdaily time step (e.g. hourly) may be required. Precipitation (time series) is the necessary input to any catchment hydrological model. That is also why hydrological models are also known as a 'precipitation–runoff' or 'rainfall–runoff' model. In the long run, no precipitation no runoff.

Other meteorological data are primarily required for evaporation and snowmelt processes. What meteorological data apart from precipitation are required for a model depends on the type of the methods used for evaporation and snowmelt computations. Simple conceptual or empirical methods for evaporation and snowmelt can be based on temperature only, e.g., the Hargreaves method for potential evapotranspiration and temperature-index approach for snowmelt (see Chapters 3 and 8). Thus, for most hydrological models precipitation and temperature are the minimum required meteorological input data. More comprehensive methods, such as the Penman-Monteith for evaporation and energy balance method for snowmelt require all of the meteorological data mentioned above (see Chapter 3).

Note that some models allow potential evaporation as a direct input, which may be obtained from other sources or computed externally. In such a case and if the catchment does not receive snow precipitation, the model may require only precipitation as meteorological input. This is also the case in most data-driven type 'rainfall–runoff' models (see Chapter 1), which are often based on precipitation and runoff data. Also, with an event-based model to simulate a short period (e.g., several hours or few days) generally with high intensity rainfall, evaporation may be considered negligible, and so the simulation may be run without temperature data.

2.1.3 Data for model calibration and validation

The most common data type for calibration/validation of a hydrological model is the river discharge. Because catchment runoff or river discharge is the key output of a hydrological model and predicting river discharge is most often its primary use, it makes good sense to use river discharge for calibration/validation. Rainfall and river discharge are the most commonly recorded hydrological data for most gauged catchments. This also makes the rainfall-runoff a perfect pair for hydrological model calibration. However, we should be clear about a possible miss-interpretation of discharge data. We usually call measured discharge, but are discharge data really measured? There are two types of discharge data, namely 'instantaneous discharge' and 'daily average discharge'. The instantaneous discharge is usually estimated through velocity and river cross-section measurements (discharge = area × average velocity). Because measuring discharge in this way in natural rivers is laborious work, the daily averaged discharge is commonly estimated from the measured water depth and the established rating curve. The instantaneous measured discharge data are usually used for establishing and updating the rating curves.

This poses an important question for hydrological model calibration: how good are the discharge data (estimated from a rating curve) against which the model is calibrated? There are several reasons why the discharge estimates can be erroneous, such as insufficient data to represent high discharge events in the rating curve, quality of discharge measurements, and hysteresis effects and lack of frequent updating of the rating curve. If we are dependent on other agencies for these data, we cannot do much about this but at the same time we will rarely have better alternatives to using the discharge data for calibration. However, if the uncertainly in the discharge data can be assessed, it can be incorporated into the model calibration and uncertainty assessment framework. More about the data uncertainty and possibility of using other types of data along with the discharge data are discussed in Chapter 9.

2.2 Data (spatial) and model resolution

With respect to representing the spatial variation, catchment models can be classified as lumped, semidistributed or distributed (Chapter 1). This is related to how the catchment area is spatially divided in the model and the size of the divided area, or more commonly the model spatial resolution. The elevation variation (topographic relief) in the catchment is represented using DEM data, which records the elevation of the area at certain spatial resolution. If a DEM has a resolution of 100 m × 100 m, for example, every 10,000 m^2 of the catchment area is represented by one elevation value (averaged over the area) in the DEM. That means that it does not show the variation of elevation within the 100 m × 100 m grid-cell (also called a pixel), and so the minimum model spatial resolution is often restricted by the DEM resolution. In a lumped model, the elevation is averaged over the whole catchment area represented in the lumped model. For example, if a catchment has an area of

$100\,km^2$, it occupies 10,000 grid-cells of $100\,m \times 100\,m$ resolution DEM. So, in a lumped model the elevations of the 10,000 grid-cells are averaged to present the elevation of the catchment. Of course, understandably it misses the existing spatial variability in the catchment, and this is the same for all spatial data represented with the same grid size.

In a semidistributed model, the catchment is divided into smaller subcatchments (based on the drainage area boundary) (Chapter 1). Then for each of the subcatchments, the elevations from the DEM pixels are averaged, which means that each subcatchment in a semidistributed model is also a lumped catchment. If the catchment of $100\,km^2$ area (same as the lumped model example above) is divided into 10 subcatchments, the average area of the subcatchments is $10\,km^2$ (note that in practice the areas of individual subcatchments vary because it is divided based on the drainage area of the tributary streams). Thus each subcatchment contains on average 1000 DEM pixels, which are averaged to represent the mean elevation of the subcatchment. So in a semidistributed model by changing the size of the subcatchments, the catchment can be represented as more distributed or less distributed. This is the same in a grid-cell based distributed model structure. Instead of dividing the catchment into subcatchments based on drainage area, in the distributed model structure the catchment is divided into grid-cells (Chapter 1). The grid-cell (or grid resolution) can be chosen small or large, but within each grid-cell the values are averaged or lumped. For example, if a distributed catchment is build using a $1\,km \times 1\,km$ grid resolution, one model pixel contains 100 DEM pixels of $100\,m \times 100\,m$ resolution, and so the elevations of the 100 pixels are averaged to represent one model pixel. This means that each grid-cell is a lumped unit in the distributed model, same as each subcatchment is a lumped unit in the semidistributed model.

Because of this some hydrological models apply some techniques to represent the subgrid or subcatchment variability. There are two most commonly used techniques to represent the subgrid or subcatchment variability. One is by dividing the subcatchments into hydrological response units (HRUs) (see Chapter 1), usually based on land use, soil and catchment slope variations. The another is by dividing the grid-cell or subcatchment areas into a number of bands or zones based on the variation in elevation, called elevation bands (zones). The division into elevation bands are particularly useful to input meteorological data (see Section 2.3.4). The HRU and elevation band divisions can be applied to all three model types (lumped, semidistributed and distributed).

If we are building a catchment model, one question we probably need to ask is what model resolution am I going to use? By model resolution, it is the grid size for the distributed structure, and size/number of subcatchments for the semidistributed structure. A more scientific question is: is there an optimal model resolution? If a coarse (spatial) resolution model serves the purpose, why make the model unnecessarily more detailed? Note also that a more detailed model does not always lead to better model performance. On this it is useful to remember the popular quote, "Everything should be made as simple as possible, but not simpler." (The quote is possibly from Einstein, possibly reworded, but apparently not fully confirmed;

see Robinson, 2018). For choosing an appropriate model resolution, a lot depends on the purpose of the model application. For example, if our aim is only to get the monthly discharge time series correct, a coarse resolution may work equally well or even better than a finer resolution model and work more efficiently. One reason why a finer resolution model does not always perform better is because it is likely less effective in calibration due to the large search space for optimal parameter sets and possible increase in equifinality (see, e.g., Beven and Freer, 2001). On the other hand, if the purpose is to use the model for forecasting floods, particularly flash floods, a coarse resolution usually cannot do a good job because by making larger spatial units we smoothen out the high intensity localized precipitation events.

2.3 Data availability and input

2.3.1 DEM, land use (land cover) and soil data

Thanks to the advancement in remote sensing technologies and data processing techniques, the DEM and land cover data are freely available as one of various remote sensing based data products. The SRTM (Shuttle Radar Topography Mission) is probably the most commonly used open access DEM data currently. The recent versions are available in two resolutions: 3 arc-second, and 1 arc-second, which are equivalent to, approximately, 90 m, and 30 m resolutions, respectively, on the line of equator. The coarse resolution version 30 arc-second (approximately 1 km) is also available. The SRTM products are available for download from more than one sources. More information about the SRTM data products and download can be found in USGS Earth Resources Observation and Science (USGS EROS Archive (website), n.d.). A recent version of the SRTM DEM can also be downloaded through the website maintained by the CGIAR—Consortium for Spatial Information (CGIAR-CSI (website), n.d.).

The size of the catchment/subcatchment, elevation variation in the catchment, and the purpose of the modelling are some of the key determining factors for choosing a DEM resolution. The 30 m resolution DEM may be found sufficient for most catchment models for general purpose hydrological simulations, except if the catchment is highly mountainous (steep elevation gradient) and a detailed representation of small streams is required. Finer resolution DEM may also be necessary to delineate flood inundation areas and to derive accurate river cross-sections. To derive accurate cross-sections of small streams and when physically based river flow methods are integrated in the catchment model, field surveyed cross-section data are usually required and recommended. Note that sometimes a small error in few DEM pixels can result in erroneous subcatchment or stream delineation. Similarly, a small error in locating the outlet point on the DEM map, particularly near a river confluence point, may result in addition or omission of a stream or subcatchment. Thus, validation of stream/catchment delineation with local field knowledge is always valuable and recommended.

Global land cover data are also available from several sources and in different resolutions. Two land cover products commonly used with hydrological models include

- Land cover from the European Space Agency (ESA) Climate Change Initiative (ESA CCI LC (website), n.d.); for product documentation, see ESA Land Cover CCI (2017); and
- Global Land Cover Characterization (GLCC) from USGS Earth Resources Observation and Science (USGSwebsite, 2018), see also USGS EROS Archive (USGS EROS Archive (website), n.d.).

For the soil data, the global soil map is available from FAO: the Digital Soil Map of the World (FAO DSMW (website), n.d.). Currently the DSMW soil data can also be downloaded from the SWAT model website (SWAT Global Data (website), n.d.).

While the quality/resolution of land cover data has recently been substantially improved, the global soil data availability is still limited. Note that some countries/regions may have locally developed land use and soil data. It is always useful to acquire such data, where possible, for the model particularly if the purpose of the model is to use locally.

Other land cover related data useful for hydrological modelling include the vegetation Leaf Area Index (LAI), Vegetation Indices, Land Cover/Dynamics, Snow Cover, and Water Reservoirs (Li et al., 2021), which are available from Moderate Resolution Imaging Spectroradiometer (MODIS) data base (MODIS Land website, 2021). Also, useful databases for modelling glacierized catchments are GLIMS Glacier Database (GLIMS (website), n.d.) and Randolph Glacier Inventory (RGI) (GLIMS RGI (website), n.d).

2.3.2 Meteorological data—precipitation

The availability of good quantity/quality precipitation data is often a major issue in catchment modelling. For modelling relatively small catchments, in situ measured precipitation data are still most commonly used. The precipitation measurement from in situ gauges is a point measurement, but we need precipitation for the entire catchment. Precipitation over a certain area or (sub)catchment, also called mean areal precipitation (MAP), can be estimated from the gauge precipitation using different methods, such as arithmetic mean, Thiessen polygon, isohyetal method and inverse distance weighting (IDW) interpolation. Precipitation can vary significantly within a catchment. There are many river catchments with immensely high rainfall variation. To give few examples, Sirisena (2020) estimated spatial variation of average annual precipitation from 3900 mm to 770 mm in the Chindwin catchment (110,926 km^2, a subcatchment of the Irrawaddy River in Myanmar), and Shrestha Zoowa (2015) based on the data from 1992 to 2007 showed spatial variation of average annual precipitation from 5647 mm to just 135 mm in the Narayani catchment (36,490 km^2, central Nepal). In the Narayani catchment example, the exceptional difference in the precipitation records within a relatively small area is due to the significant orographic effect and alignment of the neighbouring mountains (Nayava Shrestha, 2012).

With a sparse rain gauge network and/or unevenly distributed rain gauges and high rainfall variability, estimation of mean areal precipitation can be erroneous. Note also that different methods of areal precipitation/interpolation may result in different mean areal precipitation, and the difference usually increases with decreasing density of rain gauges, specially if the gauges are unevenly distributed. In the distributed model structure, the IDW method is convenient and it also offers flexibility to optimize interpolation performance by varying weighting factors through its parameters (see, e.g., Maskey, 2013; Masih et al., 2010). The interpolation performance can be assessed using jack-knife (leave-one-out) cross-validation (see Efron and Gong, 1983). The inverse distance and elevation weighting (IDEW) interpolation (e.g., used in Maskey, 2013), has been effectively used in several applications (Masih et al., 2010; Masih et al., 2011; Sirisena et al., 2018). Note that IDW or IDEW can also be used with semidistributed or lumped models to estimate (sub) catchment mean precipitation by aggregating the interpolated grid precipitation values included in each (sub)catchment.

Some catchment models may have a fixed structure to input precipitation or any other weather data, and it is always useful to know how the model uses the input data from the available gauge stations. In the SWAT hydrological model (Neitsch et al., 2011), for example, precipitation to a subcatchment is assigned from the gauge station closest to the centroid of the subcatchment, and only one station for one subcatchment. Which means that depending on the proximity of a station from the subcatchment area centroids, one station data may be used in more than one subcatchment while some stations may not be used in any subcatchments. Even a station located inside a catchment can be excluded, either because another station located in another catchment is closer than the one in the same catchment or because there are more than one stations in the same catchment. Also, if a catchment is delineated with different number of subcatchments (say for the same catchment, *n* number of catchment models with different number of subcatchments), the areal precipitation over the entire catchment in each model may be different even if exactly the same rainfall stations are kept.

One way to avoid this effect is to first generate gridded precipitation using a fixed gird resolution independent of the subcatchment structure. Then derive areal precipitation for each subcatchment from the gridded data. This way, as long as the same precipitation grids are used, the total catchment precipitation remains unchanged irrespective of the subcatchment structure or number. To input these generated precipitation data in the SWAT model, proxy stations need be assigned at the centroid of each subcatchment instead of the original stations. This technique of precipitation input solves the problem of inconsistent precipitation data input to the SWAT model in the case of a changed subcatchment structure, and it also showed to have improved the model simulation result (Masih et al., 2011).

Over the last two decades, the availability of global precipitation products has increased remarkably. Most of the recently available data products are described in Beck et al. (2017) and Mazzoleni et al. (2019). Some of the commonly used products include: TRMM (Tropical Rainfall Measuring Mission), CHIRPS (Funk et al., 2015), PERSIANN-CDR (Ashouri et al., 2015), APHRODITE (Yatagai et al., 2012) and ERA5-Land (available from Copernicus CDS (website), n.d.). These global precipitation data products may be grouped into: (1) interpolated into spatial grids from in situ gauge data, (2) satellite (remote sensing) based products, (3) reanalysis products, and (4) merged products from more than one type of data sources. These global data products have relatively coarse spatial resolution for use in a small catchment. Successes of these data products for hydrological model simulation are mixed as reported in the published literature (see, e.g., Sirisena et al., 2018; Mazzoleni et al., 2019; Nazeer et al., 2022).

Snow precipitation adds additional challenges in precipitation data input. In most cases, precipitation data records do not specify rain or snow. This distinction needs to be made in the hydrological model input, and it is a standard practice to use a threshold air temperature as the basis for the distinction (Chapter 8). If the precipitation is classified as snowfall, then the recorded precipitation depth is taken as the snow water equivalent. This separation based solely on threshold temperature is understandably an approximate method, and the threshold temperature can be a quite sensitive parameter. Note also that in the case of snowfall precipitation, because the density of new snow is difficult to measure accurately, rain precipitation and snow precipitation records may not have the same level of accuracy (see, e.g., Goodison et al., 1981, cited in Dingman, 2002). See also Chapter 8 for more discussion on issues related to snow/rainfall classification in catchment modelling.

2.3.3 Meteorological data—temperature

The availability of in situ measured temperature data is more scarce than precipitation in many river catchments, and even more so of other meteorological data, such as net radiation, humidity and wind speed. However, as temperature data are primarily used for snow accumulation and melt and (potential) evaporation, in the case of a snow free catchment, the effect of the temperature data on runoff simulation can be expected relatively less compared to the precipitation data. But in catchments with snow precipitation, the effect of precipitation and temperature are related because of the temperature threshold method used for rainfall/snowfall separation. Day-to-day variation of temperature data is generally less than precipitation data and temperature follows the diurnal cycle expect on relatively unstable weather conditions. Another favourable thing about temperature data is that it is generally more strongly elevation dependent in the same region than precipitation, which can be effectively represented through the elevation band and lapse rate (Section 2.3.4). Usually daily temperature data are reported as daily minimum, daily maximum and daily average temperatures. Because the variation of temperature over a day can be substantial, use of daily minimum and daily maximum temperatures along with the daily mean is generally preferable.

(Sub)catchment spatial average temperature can be computed using different averaging or interpolation techniques, as discussed in Section 2.3.3 for precipitation. As temperature is generally well correlated with elevation, interpolation techniques that include elevation is expected to perform better. The IDEW method can also be used for temperature. Appropriate weighting factors of the IDEW interpolation for temperature are generally different from those for precipitation interpolation.

Several global products are also available for temperature data, e.g., ERA5 (Copernicus CDS (website), n.d), CHIRTS (Climate Hazards Center (website), n. d.; Funk et al., 2019), MODIS Land Surface Temperature (MODIS Land website, 2021), APHRODITE (Yatagai et al., 2012), and CPC Global Daily Temperature (CPC (website), n.d.). The coarse spatial resolution is also the major limitation of these data products for application in relatively small catchment scale hydrological modelling.

2.3.4 Elevation bands for precipitation and temperature data

As mentioned in Section 2.2 one way to represent a subgrid or subcatchment variability is by dividing the area into a number of elevation bands. Elevation bands (or zones) are commonly used with meteorological inputs (particularly for temperature) to represent the variations within a subcatchment by means of a specified lapse rate. The lapse rate here means the rate of change of the meteorological variable value (e.g. precipitation and temperature) with respect to elevation. Elevation difference plays an important role in hydrological modelling. In mountainous catchment specially in the presence of snow/glaciers, elevation bands for temperature input play a crucial role in improving the hydrological simulation. However, note that elevation dependence of temperature is not always linear, particularly in catchments with complex topography, and the lapse rates can vary considerably on daily, monthly, seasonal and annual time scales (see, e.g., Nazeer et al., 2022).

Variation of precipitation with elevation may also be observed particularly if the areas are on the same side of the mountain range. However, the precipitation gradient with elevation is not always monotonic, which means that precipitation may increase/decrease up to certain elevation, and then reverse with further increase in elevation. Studies in the Himalayan region catchments show that elevation dependent precipitation relationship can be hardly established with a large area coverage (see, e.g., Agarwal et al., 2014, Hu et al., 2011). Immerzeel et al. (2015) discussed various difficulties in estimating precipitation in the high mountainous region. When modelling a mountainous catchment, we often have to rely on very limited data mostly from the lower elevation areas because of inaccessibility of the higher elevation regions for continuous measurements. In the largely glacierized Gilgit catchment (a subcatchment of the upper Indus basin), Nazeer et al. (2022) found that using recently available reanalysis precipitation products are a better option than to rely on lapse rate derived from few in situ stations mostly located in the lower elevations. On the other hand, in the Irrawaddy catchment (area 410,000 km^2) in Myanmar, Sirisena et al. (2018) found that interpolated precipitation data based on the IDEW method with just 19 in situ measurement stations for the entire catchment produced better runoff simulations than two global products used in the study.

References

Agarwal, A., Babel, M.S., Maskey, S., 2014. Analysis of future precipitation in the Koshi River basin, Nepal. J. Hydrol. 513, 422–434.

Ashouri, H., Hsu, K.L., Sorooshian, S., Braithwaite, D.K., Knapp, K.R., Cecil, L.D., Nelson, B.R., Prat, O.P., 2015. PERSIANN-CDR: daily precipitation climate data record from multisatellite observations for hydrological and climate studies. Bull. Am. Meteorol. Soc. 96, 69–83. https://doi.org/10.1175/BAMS-D-13-00068.1.

Beck, H.E., Vergopolan, N., Pan, M., Levizzani, V., van Dijk, A.I.J.M., Weedon, G., Brocca, L., Pappenberger, F., Huffman, G.J., Wood, E.F., 2017. Global-scale evaluation of 22 precipitation datasets using gauge observations and hydrological modelling. Hydrol. Earth Syst. Sci. Discuss. 21, 1–23. https://doi.org/10.5194/hess-2017-508.

Beven, K., Freer, J., 2001. Equifinality, data assimilation, and uncertainty estimation in mechanistic modelling of complex environmental systems using the GLUE methodology. J. Hydrol. 249, 11–29.

RGI Consortium, 2017. Randolph Glacier Inventory—A Dataset of Global Glacier Outlines: Version 6.0. Technical Report, Global Land Ice Measurements from Space. Digital Media, Colorado, USA, https://doi.org/10.7265/N5-RGI-60.

CGIAR-CSI (website), n.d. CGIAR—Consortium for Spatial Information (CGIAR-CSI). https://srtm.csi.cgiar.org/srtmdata/; last accessed on 3-November-2021.

Climate Hazards Center (website), n.d. Climate Hazards Center (CHC), University of California, Santa Barbara. https://www.chc.ucsb.edu/data/chirtsdaily, last accessed 6-November-2021.

Copernicus CDS (website), n.d. Copernicus Climate Data Store. https://cds.climate.copernicus.eu/#!/home.

CPC (website). n.d. CPC Global Daily Temperature, NOAA/OAR/ESRL PSL, Boulder, Colorado, USA. https://psl.noaa.gov/data/gridded/data.cpc.globaltemp.html, last accessed 6-November-2021.

Dingman, S.L., 2002. Physical Hydrology, second ed. Prentice Hall, New Jersey.

Efron, B., Gong, G., 1983. A leisurely look at the bootstrap, the jackknife, and cross-validation. Amer. Statist. 37 (1). https://doi.org/10.2307/2685844.

ESA CCI LC (website), n.d. European Space Agency Climate Change Initiative Land Cover Products. http://maps.elie.ucl.ac.be/CCI/viewer/download.php; last accessed 3-November-2021.

ESA Land Cover CCI, 2017. Product User Guide Version 2. Tech. Rep. Available at: maps.elie.ucl.ac.be/CCI/viewer/download/ESACCI-LC-Ph2-PUGv2_2.0.pdf.

FAO DSMW (website), n.d. FAO Digital Soil Map of the World (DSMW). https://www.fao.org/land-water/land/land-governance/land-resources-planning-toolbox/category/details/en/c/1026564/; last accessed 3-November-2021.

Funk, C., Peterson, P., Landsfeld, M., Pedreros, D., Verdin, J., Shukla, S., Husak, G., Rowland, J., Harrison, L., Hoell, A., Michaelsen, J., 2015. The climate hazards infrared precipitation with stations—a new environmental record for monitoring extremes. Sci. Data. https://doi.org/10.1038/sdata.2015.66.150066.

Funk, C., Peterson, P., Peterson, S., Shukla, S., Davenport, F., Michaelsen, J., Knapp, K.R., Landsfeld, M., Husak, G., Harrison, L., 2019. A high-resolution 1983–2016 T max climate data record based on infrared temperatures and stations by the Climate Hazard Center. J. Climate 32 (17), 5639–5658.

GLIMS (website), n.d. GLIMS: Global Land Ice Measurements from Space. https://www.glims.org/; last accessed 3-November-2021.

Goodison, B.E., Ferguson, H.L., McKay, G.A., 1981. Measurement and data analysis. In: Grey, D.M., Male, D.H. (Eds.), Handbook of snow: Principles, processes, management & use. Pergamon Press, NY.

Hu, Y., Maskey, S., Uhlenbrook, S., Zhao, H., 2011. Streamflow trends and climate linkages in the source region of the Yellow River, China. Hydrol. Process. 25, 3399–3411.

Immerzeel, W.W., Wanders, N., Lutz, A.F., Shea, J.M., Bierkens, M.F.P., 2015. Reconciling high-altitude precipitation in the upper indus basin with glacier mass balances and runoff. Hydrol. Earth Syst. Sci. 19 (11), 4673–4687.

Li, Y., Zhao, G., Shah, D., Zhao, M., Sarkar, S., Devadiga, S., Zhao, B., Zhang, S., Gao, H., 2021. NASA's MODIS/VIIRS Global Water Reservoir Product suite from moderate resolution remote sensing data. Remote Sens. (Basel) 13 (4), 565. https://doi.org/10.3390/rs13040565.

Masih, I., Uhlenbrook, S., Maskey, S., Ahmad, M.D., 2010. Regionalization of a conceptual rainfall-runoff model based on similarity of the flow duration curve: a case study from the semi-arid Karkheh Basin, Iran. J. Hydrol. 391, 188–201.

Masih, I., Maskey, S., Uhlenbrook, S., Smakhtin, V., 2011. Assessing the impact of areal precipitation input on streamflow simulations using the SWAT model. J. Am. Water Resour. Assoc. 47, 179–195. https://doi.org/10.1111/j.1752-1688.2010.00502.x.

Maskey, S., 2013. HyKit: A Tool for Grid-based Interpolation of Hydrological Variables, User's Guide (Version 1.3). IHE Delft (a copy can be obtained on request).

Mazzoleni, M., Brandimarte, L., Amaranto, A., 2019. Evaluating precipitation datasets for large-scale distributed hydrological modelling. J. Hydrol. 578. https://doi.org/10.1016/j.jhydrol.2019.124076.

MODIS Land (website), 2021. NASA MODIS Land. https://modis-land.gsfc.nasa.gov/; last accessed 3-November-2021.

Nayava Shrestha, J.L., 2012. Climates of Nepal and Their Implications. WWF Nepal, Kathmandu.

Nazeer, A., Maskey, S., Skaugen, T., McClain, M.E., 2022. Simulating the hydrological regime of the snow fed and glaciarised Gilgit Basin in the Upper Indus using global precipitation products and a data parsimonious precipitation-runoff model. Sci. Total Environ. 802. https://doi.org/10.1016/j.scitotenv.2021.149872.

Neitsch, S.L., Arnold, J.G., Kiniry, J.R., Williams, J.R. (2011). Soil and Water Assessment Tool Theoretical Documentation Version 2009. Texas Water Resources Institute. Available electronically from https://hdl.handle.net/1969.1/128050.

Robinson, A., 2018. Einstein said that—didn't he? Nature 557, 30. https://doi.org/10.1038/d41586-018-05004-4.

Shrestha Zoowa, B., 2015. Flood forecasting system for the Narayani basin in Nepal using hydrological model and weather forecast. MSc thesis, IHE Delft, the Netherlands.

Sirisena, T.A.J.G., 2020. Process based modelling of future variations in river flows and fluvial sediment supply to coasts due to climate change and human activities: data poor regions. PhD Thesis, University of Twente and IHE Delft, https://doi.org/10.3990/1.9789036550970.

Sirisena, T.A.J.G., Maskey, S., Ranasinghe, R., Babel, M.S., 2018. Effects of different precipitation inputs on streamflow simulation in the Irrawaddy River Basin, Myanmar. J. Hydrology: Regional Stud. https://doi.org/10.1016/j.ejrh.2018.10.005.

SWAT Global Data (website), n.d., SWAT Soil and Water Assessment Tool, Global Data. https://swat.tamu.edu/data; last accessed 3-November-2021.

USGS EROS Archive (website), 2018. USGS EROS Archive—Products Overview. https://www.usgs.gov/centers/eros/science/usgs-eros-archive-products-overview?qt-cience_center_objects=0#qt-science_center_objects; last accessed on 3-November-2021.

USGS EROS—GLCC (website), 2018. USGS EROS Archive—Land Cover Products—Global Land Cover Characterization (GLCC). https://www.usgs.gov/centers/eros/science/usgs-eros-archive-land-cover-products-global-land-cover-characterization-glcc?qt-science_center_objects=0#qt-science_center_objects; last accessed 3-November-2021.

Yatagai, A., Kamiguchi, K., Arakawa, O., Hamada, A., Yasutomi, N., Kitoh, A., 2012. Aphrodite constructing a long-term daily gridded precipitation dataset for Asia based on a dense network of rain gauges. Bull. Am. Meteorol. Soc. 93, 1401–1415. https://doi.org/10.1175/BAMS-D-11-00122.1.

CHAPTER

Models of evaporation and interception

3

3.1 Evaporation and evapotranspiration

As discussed in Chapter 1, evaporation is one of the four fundamental components of the catchment water balance equation; others are precipitation, runoff and storage change. Globally, around 60% of annual precipitation that falls on the terrestrial land surface turns into evaporation. On the oceans, more water evaporates than precipitates annually: $436.5 \times 10^3 \text{km}^3/\text{y}$ evaporation versus $391 \times 10^3 \text{km}^3/\text{y}$ precipitation (Oki and Kanae, 2006). The net flux of water vapour from the ocean into the atmosphere ($436.5 - 391 = 46.5 \times 10^3 \text{km}^3/\text{y}$) contributes to the total precipitation over the land surface. The proportion of evaporation relative to precipitation varies from catchment to catchment, with very high in arid and semiarid regions to low in humid regions. In fact, the ratio of precipitation to potential evaporation is used to characterize aridity of a region (see, e.g., Thornthwaite, 1948; Zomer et al., 2008; Trambauer et al., 2014a). We will discuss the difference between evaporation and potential evaporation later in the chapter. In some semiarid to arid catchments, the annual evaporation to precipitation ratio can be considerably higher than the global average, resulting in only a small fraction of precipitation as runoff. For example, in the Limpopo basin in southern Africa, Trambauer et al. (2014b) estimated the annual runoff coefficient (runoff to rainfall ratio) at around 5% (varying from 3% to 6.3% in different subbasins).

We know that besides liquid water, H_2O also exists in gaseous form as water vapour and in solid form as snow or ice. Evaporation is the process by which liquid water transforms into vapour. There are a number of things we observe in almost a daily basis that are related to the evaporation process. Few of these examples are heating a kettle of water (to make tea, for example), drying wet clothes in the open air, sticky feeling of sweat on our body or face in humid weather and seeing a wet vegetated ground surface after rainfall becoming dry the next day.

In the 'heating water kettle' example, if we observe water as it is being heated, we start seeing bubbles coming up and water vapour escaping from the surface. At this point we are basically observing water evaporating. If we continue to heat the water, while the water temperature keeps on rising the bubbles escape faster and faster. At some point the water in the kettle maintains a constant temperature, which is the boiling point. By heating the water in the kettle, we are supplying 'heat energy' to the water, and with that energy water molecules (H_2O) escape from the kettle as vapour into the air. In other words, the vapour takes energy from the water as it escapes.

Catchment Hydrological Modelling. https://doi.org/10.1016/B978-0-12-818337-3.00004-0

The more the heat energy we supply the faster the bubbles escape from the water. We also know how much energy in joules or calories is needed to evaporate a certain amount of water, which is the latent heat of vaporization (L_v). Its value varies slightly with temperature (decreases with increasing temperature), and at 15°C and one atmospheric pressure, $L_v = 2465$ kJ/kg (Monteith and Unsworth, 2013, p. 376). At the boiling point the rate of vaporization becomes such that the energy supplied to the water equals the energy lost from it due to vaporization plus any heat lost from the kettle to its surroundings, and thus the water temperature remains constant. The actual boiling point temperature depends on the pressure on the water surface.

In the 'drying wet cloths' example, it is not only the sunshine outside that controls how fast the cloths dry. If there is noticeable wind the clothes dry quicker, because the wind helps to remove the water vapour from the cloth's surface quicker and that creates a favourable condition for more water to evaporate. We will learn how the wind plays a role later in this chapter. In the example of a sweaty/sticky face, the phenomenon is related to the capacity of the air to hold water vapour. At a given temperature there is only a certain amount of water vapour the atmospheric air can hold. When the moisture in the air reaches the level that it cannot hold any more of it, the air is fully saturated. The term 'relative humidity' is used to express the amount of water vapour in the air relative to the maximum vapour it can hold (that is at saturation) at the given temperature and pressure. When the relative humidity is high the difference between the vapour density in the air and the air near the sweating skin becomes smaller, and as a result the vapour cannot transfer easily into the air and stays around the surface causing the stickiness. In less humid, dry air the sweat from our body can evaporate easily taking some heat from our body, which is how our body's cooling system works. This is also why hot temperature in a humid day is less comfortable to bear with than in a dry day. Note that while the evaporating water vapour from natural water bodies or soil takes heat energy with it, in the cooler atmosphere above, it loses that energy and condenses to water droplets in the cloud, that can eventually form rainfall. More on this topic can be found in Cutnell and Johnson (2013) and Lutgens and Tarbuck (2001). In the example of the wet vegetated ground, in addition to the radiative energy, wind and humidity, all of which play their roll, the presence of vegetation also influences evaporation through transpiration.

Transpiration is the water removed from living plants (Kent, 2013) (primarily from leaves) to the atmosphere as vapour. Transpiration is in a way same as evaporation because in the process water becomes vapour and depends on the energy supplied and weather conditions such as radiation, wind and humidity. But it is also different in the sense that it is the water that plants carry from the soil through the sap flow system from the roots through the stems and branches to the stomata in the leaves. Once the water is available in the stomata and to a little extent in pores in stems, the process of evaporation takes place again. But how the plants carry water upward from roots to stems and leaves makes a difference in when and how much water will become available in the leaves to evaporate. These processes internal to plants are not discussed in this book. Because also in transpiration the water to transform into vapour is the same thing, some hydrologists prefer to call it just evaporation without distinguishing into evaporation and transpiration (see, e.g. Ward and Robinson, 2000). However, a more commonly used term when transpiration is included is evapotranspiration (= **evapo**ration + **transpiration**). In this book, when the term 'evaporation' is used it refers to the general phenomenon of water becoming vapour, or if there is no involvement of transpiration or the distinction is not aimed or not considered necessary. When the term 'evapotranspiration' is used it refers to evaporation from soil water and/or open water as well as plant transpiration, and when the distinction is intended or considered necessary.

In the next sections we will see how the factors indicated above play a role in the formulation of evaporation equations used in catchment hydrological models. Note that evaporation including transpiration is a very rich topic covered in environmental physics and hydrometeorology. The materials covered here are a brief summary of the key concepts and methods that are of particular relevance in hydrological modelling. An in-depth coverage of the topic can be found in Brutsaert (1982) and Monteith and Unsworth (2013).

3.2 Energy balance approach for evaporation (evapotranspiration)

We have discussed a number of factors that play a role in evaporation, but whatever the factors influencing evaporation, any amount of evaporation that takes place needs energy or takes the energy from the water when it vaporizes. As we said earlier, 'evapotranspiration' is also the evaporation process (water transforms into vapour) and needs energy. The latent heat of vaporization of water is 2.465×10^6 J/kg (at 15°C and standard atmospheric pressure) means that we need 2465 kJ of energy to evaporate 1 kg of water. One Joule (J) is 1 N (force) × 1 m (distance), so the unit J is same as N.m. Therefore, if we know how much energy is available to drive evaporation, we know how much water it can evaporate, which is

$$\text{Evaporation}, E\,(\text{kg}) = \text{Energy Used}\,(\text{J})/L_v\,(\text{J/kg})$$

In hydrology we do not commonly express E in kg, but in millimetre depth of water. To get there we use the density of water, which is a wonderful value, 1000 kg/m^3. If we spread 1 kg of water over 1 m^2 area uniformly, it will have 1 mm of depth. This makes 1 kg of evaporation equivalent to 1 mm depth of evaporation from 1 m^2 of area.

In a natural environment for the evaporation to take place, the source of the energy is the Sun, or the solar radiation. We know that solar radiation is often expressed in W/m^2, and Watt is Joule/second, we can also write

$$\text{Evaporation}, E\,(\text{mm/s}) = \text{Energy Used}\left(\text{W/m}^2\right)/L_v\,(\text{J/kg})$$

See Box 3.1 for an example to convert Joule per time (or Watt) energy to an equivalent evaporation depth per time.

Because of this, one obvious approach to estimate evaporation is using the energy balance. Similar to the water balance in which we carry out a balance between all

Box 3.1

The Earth's energy balance suggests that of the 340 W/m^2 of daily average incoming solar radiation at the top of the atmosphere, about 185 W/m^2 reaches the earth's surface and about 85 W/m^2 of that is used as latent heat (these values are based on Wild et al., 2013). How much evaporation in mm/day is equivalent to 85 W/m^2 (average throughout day and night) with $L_v = 2.465 \times 10^6$ J/kg?

Solution

Total energy used as latent heat in a day = 85 × (60 × 60 × 24) J/d/m^2

$$E = 85 \times 86400/(2.465 \times 10^6) \approx 3\,\text{mm/d}$$

inflow and outflow sources of water in a control volume, in energy balance we do the same for all incoming and outgoing sources of energy on an object or a unit area surface.

So, what are the components of the energy balance equation over an area on earth's surface? First and foremost, there are two types of radiation, namely shortwave (incoming and outgoing), and longwave (incoming and outgoing). The incoming shortwave radiation is the solar radiation (emitted from the sun) that reaches the earth's surface, and the outgoing radiation is part of the incoming solar radiation reflected by the receiving surface. They are called shortwave because the solar radiation is composed of the smaller wave lengths in the total spectrum relevant to the Earth's energy budget. To recall, Wien's displacement law states that the wavelength of maximum intensity is inversely proportional to the temperature of the radiating surface. The longwave radiations are due to the emission from the earth itself: the outgoing (upward) longwave radiation from the earth's surface to the atmosphere and the incoming (downward) from the atmosphere to earth's surface. Because earth's temperature is much cooler than that of the surface of the sun, spectral wavelengths of earth's emission are much larger than those of the solar radiation. To compare, the average temperature of the surface of the sun is roughly around 6000 K, and average temperature of earth's surface is around 288 K (see, e.g. Lutgens and Tarbuck, 2001).

The other components of the energy balance are the ground heat flux (G), which is due to heat energy transfer through conduction to the subsurface soil, the sensible heat flux (H) due to molecular transfer of heat to the air above the surface, and the latent heat flux due to vapour transfer in the evaporation process (L_vE). For a snow or ice surface, the required latent heat component is somewhat larger for both melting and evaporating the frozen water, which is called sublimation. We will write the energy balance equation for the snow/ice surface in Chapter 8. Thus, we can write the energy balance equation as

$$R_S^{\downarrow} - R_S^{\uparrow} + R_L^{\downarrow} - R_L^{\uparrow} = G + H + L_vE \tag{3.1}$$

On the left-hand side, the upward and downward radiation terms are indicated by an arrow for both shortwave and longwave radiation. Commonly, downward terms (toward the surface) are taken positive, and upward terms (away from the surface) are taken negative. The net allwave radiation balance is conveniently represented by R_N, that is

$$R_N = R_S^{\downarrow} - R_S^{\uparrow} + R_L^{\downarrow} - R_L^{\uparrow} \quad (3.2)$$

Thus, the latent heat flux (evaporation) is given by

$$L_v E = R_N - G - H \quad (3.3)$$

In this way, we need to know three quantities (R_N, G and H) to be able to estimate evaporation. How do we obtain or calculate these terms? Although strictly speaking it is possible to measure the shortwave and longwave radiation, incoming and outgoing, but in practice radiation measurements are much more scarce than, e.g., precipitation and temperature. In the absence of direct measurements, these radiations are often estimated from measurements of other parameters and relationships.

The incoming shortwave radiation can be estimated from the known solar constant (the solar radiation at the top of the atmosphere) with adjustment for latitude and position of the sun around the Equator—indicated by the day of the year. The solar constant, R_{so} is 1367 W/m^2 but given the earth's rotation, the amount of solar radiation received at the surface varies with the time of the day. On the earth's surface, the incoming solar radiation is also dependent on the altitude of the location, because at a higher altitude the solar radiation needs to penetrate through thinner atmosphere than at lower altitude. However, the major factor that affects the amount of solar radiation to reach the earth's surface at any given location and time is the atmospheric transmissivity with and without clouds. A thick cloud cover can reflect a large part of the solar radiation. The atmospheric transmissivity at sea level on a cloud-free day is about 0.84, under thick haze it is about 0.6, and may reduce to nearly zero under thick clouds (Cuffey and Paterson, 2010, p. 143). However, even under dense cloud cover, a part of the extraterrestrial radiation may reach the earth's surface as diffuse sky radiation (Allen et al., 1998). In practical applications, the ratio of actual measured sunshine hours to the maximum possible cloud-free sunshine hours on any day at any given location is often used as a proxy for the cloud effects on incoming solar radiation, $R_S(\downarrow)$. A detailed procedure with an example to estimate incoming shortwave radiation from the extraterrestrial solar radiation is presented in the FAO-56 report (Allen et al., 1998).

The outgoing shortwave radiation is the fraction of the incoming (downward) shortwave radiation at the surface that is reflected by the surface. It can be estimated from the albedo (α) or reflectivity of the surface. This means that the net shortwave radiation ($R_{S,N}$) can be expressed as

$$R_{S,N} = (1 - \alpha) R_S^{\downarrow} \quad (3.4)$$

The upward longwave radiation can be estimated from the temperature of the earth's surface and downward longwave radiation from the effective radiative temperature of the atmosphere using the Stefan-Boltzmann Law of Radiation, which states that the radiative energy emitted, E_m (in W/m^2), by a surface is proportional to its temperature (in Kelvin) to the power of four. That is, the radiant energy emitted

$$E_m\ (\text{in W/m}^2) = \text{Emissivity}\ (\varepsilon) \times \text{a constant}\ (\sigma) \times T^4$$

where T is temperature (in K) and the Stefan-Boltzmann constant, $\sigma = 5.67 \times 10^{-8}\,\text{W m}^{-2}\text{K}^{-4}$.

Thus, the downward, upward and net longwave radiations can be expressed as

$$R_L^{\downarrow} = \varepsilon_{atm}\sigma T_a^{\ 4} \tag{3.5}$$

$$R_L^{\uparrow} = \varepsilon_s\sigma T_s^{\ 4} \tag{3.6}$$

$$R_{L,N} = \sigma\left(\varepsilon_{atm}T_a^{\ 4} - \varepsilon_s T_s^{\ 4}\right) \tag{3.7}$$

where ε_{atm} and ε_s are the emissivity values for the atmosphere and the earth's surface, and T_{a} (in K) is the air temperature (usually specified at 2 m height) and T_{s} (in K) is the surface temperature.

The other two terms in the energy balance—the ground heat flux G and the sensible heat flux due to air turbulence above the surface H—refer to the transfer of heat energy through the soil and air, respectively. The ground heat flux G is the heat energy transfer from the surface to the soil underneath through conduction. It is calculated from the temperature gradient in the soil layer, that is the difference in temperature divided by the layer thickness, and the soil thermal conductivity (k_s). The relation is known as Fourier's law of heat conduction:

$$G = -k_s\frac{\partial T}{\partial z} \tag{3.8}$$

where G is in W/m^2, T in °C, z in m, and so k_s in W/m/°C. The flux moves from higher temperature to lower temperature. Thus, the negative sign here indicates that if the heat flow is from the surface downward into the soil below, the flux G is positive. Because z is taken positive downward, when the temperature at the surface is higher than at a depth Δz below, the temperature difference, ΔT and so the temperature gradient is negative, and as a result the flux is positive. Which is usually the case in the day time when the ground surface heats up due to the solar radiation, and it is generally the opposite in the night.

The sensible heat transfer is due to the turbulent transfer through the air, and can be represented in the same manner as G with the temperature gradient, that is

$$H = -\rho_a c_p k_H\frac{\partial T}{\partial z} \tag{3.9}$$

Here, H is in W/m^2, ρ_{a} is the density of air in kg/m^3, c_p is the specific heat capacity of air at constant pressure in J/kg/°C, T in °C, z in m, and so k_H, the thermal diffusivity of air, in m^2/s. Note the difference in units of k_s in Eq. (3.8) and k_H in Eq. (3.9). In the two equations, G and H have the same unit, so the unit of k_s is same as the unit of ($\rho_{\text{a}}\ c_p\ k_H$). Because ρ_{a} and c_p are typically known for atmospheric air, it is useful to define Eq. (3.9) in this way. The negative sign in the expression for H has the same purpose as for G. Both G and H take the sign (negative or positive) opposite to that of the temperature gradient. In the case of H, heat transfer through the air, z is taken positive upward. That means, if T at a certain height from the surface is lower than at the surface, the temperature gradient is negative, and so H is positive. Thus, heat transfer from surface to the atmosphere upward is taken as positive. In the day time when the ground warms up, the heat transfer is from surface to the atmosphere, and it generally reverses in the night.

In theory the ground heat flux can be estimated knowing the temperature difference in the soil layer (e.g. by measurement) and the thermal conductivity of the soil and moisture composition. However, in practice, computing G is limited due to lack of full knowledge of the soil composition, its related properties and spatial heterogeneity, and lack of soil temperature measurements. Moreover, it is also much harder to measure or estimate the sensible heat flux with the seemingly simple relationship shown above. Because of these difficulties, application of the energy balance method of evaporation estimation is usually limited.

One approximation may be sought by using a ratio of the sensible heat (H) to latent heat (L_vE) in place of estimating H directly. Thus, making use of the Bowen ratio $\beta = H / (L_vE)$, we obtain

$$L_vE = R_N - G - \beta(L_vE) \tag{3.10}$$

$$L_vE = \frac{R_N - G}{1+\beta} \tag{3.11}$$

The Bowen ratio (after Bowen, 1926) may be estimated from the difference in saturated vapour pressure (e_s) and actual vapour pressure in the air at a specified height (e_a), and the difference in the temperature of the evaporating surface (T_s) and the mean air temperature (T_a). As such it is given by the equation (Penman, 1956; Brutsaert, 2005)

$$\beta = \frac{\gamma(T_s - T_a)}{(e_s - e_a)} \tag{3.12}$$

where γ is the psychometric constant given by

$$\gamma = \frac{c_p p}{L_v \varepsilon} \tag{3.13}$$

where p is the atmospheric pressure and $\varepsilon = 0.622$ is the ratio of molecular weight of water vapour to that of dry air. For a temperature of 20°C and an atmospheric pressure of 101.3 kPa, $\gamma = 0.067$ kPa/K is obtained (see Brutsaert, 2005).

3.3 Penman and Penman-Monteith equations

The Penman equation (Penman, 1948) and Penman-Monteith equation (Monteith, 1965) are probably the most commonly used evaporation methods in hydrological modelling. Next to the energy balance method described in Section 3.2, the Penman and Penman-Monteith are among the comprehensive physically based methods and are based on the combination of energy balance and aerodynamic approaches. The starting point of the Penman equation is also the energy balance, but it attempts to quantify the energy fluxes using the physics of heat and mass transfer. The energy balance equation is written here without the ground heat flux (G), which is not included in the Penman equation.

$$R_N = H + L_v E \tag{3.14}$$

The net radiative energy (R_N) is balanced by the latent heat flux ($L_v E$) and the sensible heat flux (H).

In Section 3.2, the sensible heat flux was expressed using the temperature gradient. In the same manner, the latent heat flux can be expressed using the gradient in density of moist air (ρ_v) as (Oke, 1987, p. 65)

$$L_v E = -L_v k_v \frac{\partial \rho_v}{\partial z} \tag{3.15}$$

where k_v is the eddy diffusivity of water vapour (m^2/s).

One convenient analogy used to describe heat and vapour transfer is the analogy of Ohm's law of electric current through a piece of material, in which the current (I) is equal to the potential difference, or voltage (V), divided by the electrical resistance (R) of the material ($I = V / R$). Using this analogy, the vapour and heat transfer fluxes can be expressed as

$$\text{Transfer flux } (H \text{ or } E) = \text{Potential difference/Resistance}$$

Comparing with the gradient-based expressions (Eq. 3.9 and 3.15), we can see that potential differences come from the temperature difference between the surface and the air for the heat transfer and the vapour density (or humidity) difference for the latent heat transfer. Note that vapour pressure is also often used in place of vapour density as they are closely related. Vapour pressure (e) is the part of the total atmospheric pressure exerted by the water vapour present in the air. The resistance is the aerodynamic resistance (r_a) near the surface. The aerodynamic resistance may also be differentiated into the resistance to heat transfer (r_H) and the resistance to vapour transfer (r_v). The common unit of r_a, r_H and r_v is s/m.

Furthermore, the concept of "wet-bulb temperature" is useful here, in which the temperature is measured with a sensor that is covered with a wet material (e.g. wet fabric). Such a measurement would record lower temperature than a sensor without a wet cover, because the wet cover loses energy as latent heat by evaporation. If this system is in an adiabatic condition (that is the net energy exchange is zero), the sum of the latent heat and sensible heat must equal zero. To understand this, assume a closed jar partly filled with water and perfectly insulated to keep it in adiabatic condition. When the water starts evaporating, the air in the jar gains energy (latent heat) through the movement of water molecules and as a result the vapour pressure in the air increases. Because it is in adiabatic condition, the increase in energy will be balanced by the cooling of the air (as sensible heat). This process continues until the air in the jar is fully saturated.

Suppose at the initial state, the air in the jar has temperature T and vapour pressure e, such that $e < e_s(T)$. Where, $e_s(T)$ is the saturation vapour pressure at the given temperature. Let us say, this condition (T, e) is shown in Fig. 3.1 (bottom panel) by point A. Note that in Fig. 3.1 the mixing ratio (top panel) is shown for additional information only. At the same temperature T, the air would saturate with the saturated vapour pressure $e_s(T)$ shown by point C. On the other hand, for the air to saturate with the same vapour pressure as e, the temperature needs to drop to T_d, which is the dew point temperature. In the case of the wet-bulb analogy, when there is enough water to evaporate and if the system is in adiabatic condition (meaning no radiative energy loss or gain, and the latent heat is balanced by the sensible heat transfer), it will reach an equilibrium at temperature T_w (wet-bulb temperature) when the air fully saturates somewhere between T_d and T as shown by point D in the figure. In this condition, and using the resistance formulation, the heat lost through the latent heat flux (evaporation) and the balancing sensible heat flux can be expressed by (see, e.g., Monteith and Unsworth, 2013)

$$L_vE = \frac{\rho_a c_p}{\gamma}\left(\frac{e_s(T_w) - e}{r_v}\right) \tag{3.16}$$

$$H = \rho_a c_p\left(\frac{T_w - T}{r_H}\right) \tag{3.17}$$

Finally, to derive the Penman equation, the evaporating surface is assumed to be a water surface or a fully wet surface, so that the vapour pressure at the surface is assumed to be the saturation vapour pressure. Thus, Eqs. (3.16) and (3.17) are rewritten substituting T_w by T_0, where T_0 is the temperature of the surface, and assuming $r_v = r_H = r_a$. That is

$$L_vE = \frac{\rho_a c_p}{\gamma}\left(\frac{e_s(T_0) - e}{r_a}\right) \tag{3.18}$$

$$H = \rho_a c_p\left(\frac{T_0 - T}{r_a}\right) \tag{3.19}$$

The latter equation can be rearranged as

$$T_0 - T = \frac{r_a H}{\rho_a c_p} \tag{3.20}$$

The important part of the Penman and Penman-Monteith equations is to exclude the surface temperature because in practice it is difficult to measure and such measurements are not part of standard meteorological observations. To get that, let us say the slope of the saturation vapour pressure vs temperature curve (Fig. 3.1) is s_e. Then, assuming a linear approximation of the vapour pressure curve between T_w and T, we can write

$$e_s(T_w) = e_s(T) - s_e(T - T_w) \tag{3.21}$$

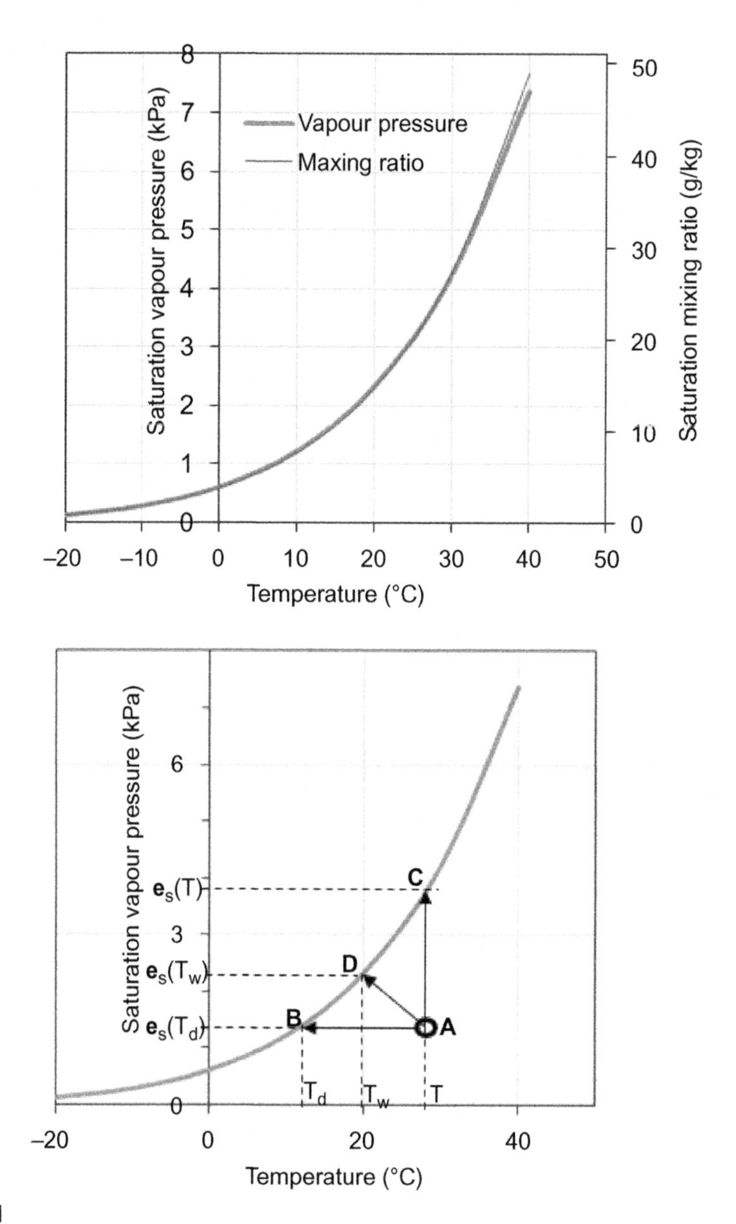

FIG. 3.1

Top: Temperature versus saturation (or equilibrium) vapour pressure (left y-axis) and saturation mixing ratio (g/kg), which is the mass of vapour per kg mass of dry air (right y-axis). (The maxing ratio is shown for additional information only). Bottom: Illustration of wet-bulb temperature (T_w), dew-point temperature (T_d) and saturation vapour pressures at T, T_w and T_d. Values of the saturation vapour pressure are from Oke (1987) (p. 394), which are identical with the values estimated from Eq. (3.38) up to two decimal points in kPa. Values of the saturation mixing ratio are estimated from the saturation vapour pressure at mean sea level pressure, i.e. $m_{r,\text{sat}}(T) = 0.622e_s(T)/(101.3 - e_s(T))$.

Assuming the water or wet surface assumption as before, $T_w = T_0$ is used to yield

$$\begin{aligned} e_s(T_0) &= e_s(T) - s_e(T - T_0) \\ &= e_s(T) + s_e(T_0 - T) \end{aligned} \tag{3.22}$$

This allows to replace $e_s(T_0)$ in Eq. (3.18) to yield

$$L_v E = \frac{\rho_a c_p (e_s(T) + s_e(T_0 - T) - e)}{\gamma r_a} \tag{3.23}$$

Then replacing $(T_0 - T)$ from Eq. (3.20), we obtain

$$L_v E = \frac{\rho_a c_p \left(e_s(T) - e + s_e \frac{r_a H}{\rho c_p} \right)}{\gamma r_a} \tag{3.24}$$

At this point it is also convenient to write $e = e_a$ and $e_s(T) = e_s$ to mean the actual vapour pressure at a given temperature and the saturation vapour pressure at the same temperature, respectively. Also, representing $H = R_N - L_v E$ from Eq. (3.14) for non-adiabatic conditions, we obtain

$$L_v E = \frac{\rho_a c_p \left(e_s - e_a + s_e \frac{r_a}{\rho_a c_p} (R_N - L_v E) \right)}{\gamma r_a} \tag{3.25}$$

Rearranging, the **Penman equation** is written as

$$L_v E = \frac{s_e R_N}{s_e + \gamma} + \frac{\rho_a c_p (e_s - e_a)}{(s_e + \gamma) r_a} \tag{3.26}$$

Note that the original Penman equation (Penman, 1948) uses a wind function in place of the aerodynamic resistance, which is discussed separately later in this chapter.

In case the ground heat flux is included in the energy balance, Eq. (3.26) can be written as

$$L_v E = \frac{s_e (R_N - G)}{s_e + \gamma} + \frac{\rho_a c_p (e_s - e_a)}{(s_e + \gamma) r_a} \tag{3.27}$$

3.3.1 Penman-Monteith equation

Although the Penman method given by Eq. (3.26) is commonly used to estimate evaporation or evapotranspiration from open water or fully wet surfaces, it can be used to other surfaces, e.g. partly wet or vegetated, as long as the resistance terms are adjusted to represent the type of the surfaces (Monteith and Unsworth, 2013). In a vegetated surface, clearly an additional resistance to water vapour transfer (r_v) can be observed due to the stomata in the plant leaves, r_s. Thus, the latent heat flux would be

$$L_v E = \frac{\rho_a c_p}{\gamma} \left(\frac{e_s(T_0) - e}{r_a + r_s} \right) \tag{3.28}$$

Using this form of the latent heat equation, which has $r_a + r_s$ as total resistance in the above derivation yields the Penman-Monteith equation (Monteith, 1965) as

$$L_v E = \frac{s_e R_N + \rho_a c_p (e_s - e_a)(1/r_a)}{s_e + \gamma(1 + r_s/r_a)} \tag{3.29}$$

Similar to that in the Penman equation, if the ground heat flux is also considered, the Penman-Monteith equation becomes

$$L_v E = \frac{s_e (R_N - G) + \rho_a c_p (e_s - e_a)(1/r_a)}{s_e + \gamma(1 + r_s/r_a)} \tag{3.30}$$

The Penman-Monteith equation is also presented with a new resistance term called surface resistance (r_c) to apply it on different types of vegetation and soil surfaces (Monteith and Unsworth, 2013). The surface resistance replaces the stomata resistance, that is

$$L_v E = \frac{s_e (R_N - G) + \rho_a c_p (e_s - e_a)(1/r_a)}{s_e + \gamma(1 + r_c/r_a)} \tag{3.31}$$

If you are first time attempting to use Penman-Monteith (or Penman) equation, it is worth checking the units of the equation parameters and how they make up the unit of the evaporation flux as described in Box 3.2.

3.3.2 Penman equation with the wind function

Another form of the Penman equation uses directly a function of wind speed, $f(u_2)$, instead of the aerodynamic resistance.

Box 3.2 Units of Penman-Monteith equation parameters

In the application of the Penman-Monteith equation, one of the confusions may arise because it involves a number of parameters with different units. So, if you are a first-time user of the equation, it is worth looking into the units carefully. Let's take Eq. (3.29). The right-hand side of this equation can be separated into three blocks, as (A+B) / C. Then

$$\text{Unit of A} = \text{Unit of } (s_e R_N) = (\text{kP/K})\,(\text{W/m}^2)$$

$$\text{Unit of B} = \text{Unit of } \rho_a c_p (e_s - e_a)\,(1/\text{r}_\text{a}) = (\text{kg/m}^3)\,(\text{J/kg/K})\,(\text{kP})\,(\text{m/s})$$

$$= (\text{J/s/m}^2)(\text{kP/K}) = (\text{W/m}^2)(\text{kP/K})$$

$$\text{Unit of C} = \text{Unit of } s_e \text{ or unit of } \gamma = \text{kPa/K}.$$

Thus, the unit of (A + B) / C = (kPa/K) (W/m^2) / (kPa/K) = W/m^2

In the left-hand side, we have $L_v E$ whose unit must also be W/m^2. We know the unit of L_v is J/kg.

Thus, the unit of E = Unit of (1/L_v) (W/m^2) = (kg/J) (W/m^2) = (kg/J) (J/s/m^2) = kg/s/m^2 (or mm/s)

Note that the parameters used in the equation (e.g. T, R_h, etc.) are averaged over a certain period. Most applications in hydrological modelling use a daily or sometimes hourly time steps. Time conversions must be applied to obtain the desired evaporation rate units (e.g. from mm/s to mm/day or mm/h).

$$E=\frac{s_e}{L_v(s_e+\gamma)}R_N+\frac{\gamma}{(s_e+\gamma)}f(u_2)(e_s-e_a) \quad (3.32)$$

In the Penman (1948) formulation, the wind function is defined empirically as a linear function of the wind speed at 2 m height in the form $f(u_2)=C(A+Bu_2)$. In his original equation, the wind speed is in miles per day, vapour pressure deficit (e_s-e_a) in mm of Hg and the evaporation in mm/day. With these units, the wind function parameters in Penman (1948) are $C=0.35$, $A=1.0$ and $B=0.0098$. Penman (1956) recognized that the evaporation estimated by these parameters was generally high, and so proposed a revised value for parameter $A=0.5$. To use the wind function in SI units, two conversions are required: the wind speed from miles per day to m/s and vapour pressure deficit from mm of Hg to kPa. With 1 miles per day = 1609/86400 m/s and 1 mmHg pressure = 101.3/760 kPa, the wind function parameters become $C=2.626$, $A=0.5$, and $B=0.526$. This wind function when presented in the form $A+Bu_2$ becomes (see e.g. McMahon et al., 2013)

$$f(u_2)=1.313+1.381u_2 \quad (3.33)$$

Note that Penman (1956) presented the parameter B rounding off to 0.01 (instead of 0.0098) which would be 0.537 after the unit conversion, and the wind function becomes

$$f(u_2)=1.313+1.41u_2 \quad (3.34)$$

In the Penman equation (3.26), the wind function is related to the aerodynamic resistance r_a. Comparing the two equations (Eqs. 3.26 and 3.32), we obtain

$$\left(\frac{86400}{L_v}\right)\frac{\rho_a c_p}{r_a}=\gamma f(u_2) \quad (3.35)$$

Replacing the psychometric constant by $0.665\times10^{-3}p$ and rearranging yields

$$r_a=\frac{86400}{L_v}\frac{1000\rho_a c_p}{0.665pf(u_2)} \quad (3.36)$$

We see that relating r_a with the Penman's $f(u_2)$ function is tricky because the relationship is not a constant, but varies with pressure, temperature and humidity. The bottom line about the wind function is that it is an empirical function, and in principle it should be calibrated for its application in specific conditions. We will discuss about model calibration and uncertainty in Chapter 9. Efforts to define a r_a to u_2 relationship assume certain (standard) values of air density, pressure and temperature. For the Penman (1956) wind function Eq. (3.34) ($C=2.626$, $A=0.5$ and $B=0.54$), and taking atmospheric pressure 101.3 kPa (at mean sea level), air density 1.23 kg/m^3 and the constants $c_p=1004$ J/kg/°C and $L_v=2.465\times10^6$ J/kg, following relationship can be found:

$$r_a=\frac{245}{0.5+0.54u_2} \quad (3.37)$$

where u_2 in m/s and r_a in s/m.

3.4 Estimation of Penman and Penman-Monteith equation parameters

The final forms of the Penman and Penman-Monteith equations are quite compact in terms of the number of inputs or parameters they have. To apply these equations, depending on the available data, for some parameters we usually/often need to rely on a combination of physics-based and empirical relationships. The required meteorological variables, relationships and thermodynamic/physical properties of water and air are described below. Most of the relationships described here are also used in the FAO-56 guidelines for estimating crop water requirement (Allen et al., 1998).

Following meteorological variable data are necessary for the Penman and Penman-Monteith methods, which are averaged over the required time period (e.g. hourly, daily):

- Net radiation, R_N
- Relative humidity at 2 m height, R_h
- Wind speed at 2 m height, u_2
- Air temperature at 2 m height, T; preferably also minimum (T_{min}) and maximum (T_{max}) over the period.

3.4.1 Latent heat of vaporization (L_v) and specific heat capacity of air at constant pressure (c_p)

These are often taken as constant values: $L_v = 2.45$ MJ/kg (at 20°C) and for average atmospheric conditions, $c_p = 1.013 \times 10^{-3}$ MJ/kg/°C (Allen et al., 1998).

3.4.2 Saturation vapour pressure and actual vapour pressure

The vapour pressure deficit ($e_s - e_a$) can be computed from relative humidity (R_h), because $R_h = e_a/e_s$, and the known relationship for e_s for a given temperature at constant pressure as represented in Fig. 3.1. The expression for e_s applicable over a water surface derived by Murray (1967) based on the equation by Tetens (1930) (cited in Murray, 1967 and Monteith and Unsworth, 2013), also used in Allen et al. (1998) and Chow et al. (1988), is

$$e_s = 0.6108 \exp\left(\frac{17.27T}{237.3 + T}\right) \tag{3.38}$$

$$e_a = e_s R_h \tag{3.39}$$

where e_s and e_a are in kPa and air temperature T is in °C. Note that the constant 0.6108 is the e_s at 0°C in kPa.

3.4.3 Slope of the vapour pressure curve (s_e)

It can be derived by differentiating the above saturation vapour pressure equation with respect to T. That is (see also Box 3.3)

$$s_e = \frac{de_s}{dT} = \frac{4098 e_s}{(237.3 + T)^2} \tag{3.40}$$

where s_e in kPa/°C, e_s in kPa and T in °C.

3.4.4 Psychometric constant (γ)

It is defined by Eq. (3.13) and its value for a specified pressure and temperature was also discussed earlier. It may also be approximated as a function of pressure assuming constant temperature for simplicity. Using the values of c_p and L_v as above and taking $\varepsilon = 0.622$, we obtain

$$\gamma = 0.665 \times 10^{-3} p \tag{3.41}$$

where p is in kPa and γ is in kPa/°C.

Box 3.3 Derivation of Eq. (3.40) from Eq. (3.38)

Let us rewrite Eq. (3.38) as

$$e_s = c\exp\left(\frac{aT}{b+T}\right)$$

where the constants, $c = 0.6108$, $a = 17.27$, $b = 237.3$. Then

$$\begin{aligned} s_e = \frac{de_s}{dT} &= c\exp\left(\frac{aT}{b+T}\right)\frac{d\left(\frac{aT}{b+T}\right)}{dT} \\ &= e_s a\left(T\frac{d(1/(b+T))}{dT} + \frac{1}{b+T}\frac{dT}{dT}\right) \\ &= e_s a\left(\frac{-T}{(b+T)^2} + \frac{1}{b+T}\right) \\ &= e_s\frac{ab}{(b+T)^2} \end{aligned}$$

Finally substituting the values of a and b

$$s_e = \frac{4098 e_s}{(237.3 + T)^2}$$

3.4.5 Atmospheric pressure (p)

At mean sea level, the average atmospheric pressure is 101,325 Pa (usually expressed as 101.3 kPa), which is equivalent to the pressure exerted by 760 mm column of mercury (Hg) or 10.34 m column of water (Streeter and Wylie, 1983). 1 Pascal is 1 N/m^2. The atmospheric pressure varies with elevation, which affects the atmospheric column height and air density. If the local atmospheric pressure is not given/known, it can be approximated from the elevation of the location (z) and standard mean sea level pressure (101.3 kPa) using the relation (see, e.g. Allen et al., 1998)

$$p(z) = 101.3\left(1 - \frac{0.0065z}{293}\right)^{5.26} \tag{3.42}$$

where $p(z)$ is in kPa and elevation z is in m with the datum at mean sea level.

3.4.6 Air density (ρ_a)

Air density varies with pressure and temperature and also depends on how much water vapour is present in the air because it is the sum of the density of water vapour and dry air ($\rho_a = \rho_w + \rho_d$). It can be represented by the following relationship (see, for derivation steps, Brutsaert, 2005, Wallace and Peter, 2006):

$$\rho_a = \frac{1}{R_d T}(p - 0.378e) \tag{3.43}$$

where ρ_a is in kg/m^3, p and e are in Pa, T is in K and R_d is the specific gas constant of dry air in J/kg/K. Taking the value of $R_d = 287$ J/kg/K, and expressing p and e in kPa and T in °C, we obtain

$$\rho_a = \frac{p - 0.378e}{0.287(273 + T)} \tag{3.44}$$

3.4.7 Aerodynamic resistance (r_a) and surface resistances (r_c)

Aerodynamic resistance (r_a) is related to wind speed but very difficult to measure. To use with the Penman method for open water, Eq. (3.37) may be used, which is equivalent to the wind function used by Penman (1948). For an average wind speed of 2 m/s, Eq. (3.37) yields $r_a = 155$ s/m.

However, different versions of the relationships can be found in the literature. One of those is proposed by Thom and Oliver (1977) (cited in De Bruin, 1982; Chin, 2011) for open water surface, which has the possibility to specify the height of wind measurement (z_w) and aerodynamic roughness height (z_0):

$$r_a = \frac{4.72 \ln\left(\frac{z_w}{z_0}\right)^2}{1 + 0.54u_2} \tag{3.45}$$

where z_w and z_0 are in m, u_2 in m/s and r_a in s/m.

To use the Penman-Monteith equation as reference evapotranspiration (ET_0) (see Section 3.5), the r_a and r_c values derived by Allen et al. (1998) can be used, which uses a "reference surface" defined as a hypothetical surface area covered with grass of height = 0.12 m, surface resistance (a bulk resistance that also includes the effect of stomata resistance) $r_c = 70$ s/m, and albedo = 0.23. The aerodynamic resistance is given by

$$r_a = \frac{208}{u_2} \tag{3.46}$$

where u_2 is in m/s and r_a in s/m.

Few examples of minimum r_c values presented in Monteith and Unsworth (2013) based on Kelliher et al. (1995) vary from 30 s/m for cereal crops to 80 s/m for tropical rain forest. The same for the temperate grassland and temperate deciduous forest are 60 s/m and 50 s/m, respectively.

3.5 Potential, actual and reference evaporation (evapotranspiration)

The concepts of 'actual evaporation' and 'potential evaporation' are in some way ambiguous. The same can be said about 'actual evapotranspiration' (AET) and 'potential evapotranspiration' (PET). Besides actual and potential, there is also 'reference' evapotranspiration, although the definition of the latter is more consistent as long as the 'reference' surface is clearly defined. The term 'potential evapotranspiration' was first used by Thornthwaite (1948) to distinguish evapotranspiration from an area that is actually taking place (the 'actual evapotranspiration') to the hypothetical evaporation that would have taken place if the area is under "ideal conditions of soil moisture and vegetation". The "ideal" condition primarily refers to the ample availability of water in the soil for the type of vegetation. The concept was subsequently used by Penman (see Penman, 1956), although he preferred/used 'potential transpiration' over 'potential evapotranspiration'. One source of ambiguity arises from the fact that in the energy balance approach the water sufficiency/deficiency condition does not come into the picture directly. If the evaporation is determined from the residual of the energy inflow/outflow terms ($L_v E = R_N - H - G$), it should represent the 'actual evaporation' as long as the other terms of the energy balance are accurately quantified. However, in practice the evaporation estimates from the energy balance method as well as from the Penman or Penman-Monteith method are often treated as potential evaporation (evapotranspiration) than actual evaporation (evapotranspiration), in most hydrological models. The reason is primarily because it is difficult to accurately measure or estimate all the input data and parameters for these methods, and that may lead to the situation when estimated evaporation is more than available moisture during a given time period. Therefore, despite some of the ambiguities (see more in Brutsaert, 2005; Ward and Robinson, 2000), it is a common practice in catchment modelling to consider evaporation estimates from the energy-balance or, more generally, methods driven by climatic input (including Penman and Penman-Monteith) as 'potential evapotranspiration' or an 'upper limit' to 'actual evaporation'. The actual evaporation is then estimated separately from different surfaces, keeping the total over an area less or equal to the potential. Thus, the practical assumption behind the use of PET as an upper limit for AET is that the estimated PET is without water shortage.

The concept of the "reference evapotranspiration" was recommended in the FAO-24 guidelines for computing crop water requirements (Doorenbos and Pruitt, 1977) and later revised in FAO-56 (Allen et al., 1998). In the "reference evapotranspiration" a hypothetical vegetation surface is defined as a reference surface with specified characteristics (see Section 3.4.7). Furthermore, FAO-56 recommends the Penman-Monteith equation to estimate reference evapotranspiration (ET_{ref}) for the reference surface. The idea is that once the ET_{ref} is established for the reference surface in a given climatic condition, any other vegetation surface can be compared with the reference surface to determine its water requirement (as evapotranspiration) for different agronomic and water management conditions. As such, a factor called a crop coefficient is applied to the ET_{ref} to estimate actual evapotranspiration for the crop type and growth cycle stage of interest.

3.6 Simplified methods for potential evapotranspiration

As we discussed earlier, the energy balance as well as the Penman and Penman-Monteith methods are data demanding. Where climate data such as radiation, humidity, temperature and wind speed are available, Penman-Monteith has been almost like a standard method for evaporation in hydrological modelling. The popularity of the Penman-Monteith method has certainly increased since FAO-56 has recommended it as the standard method for reference evapotranspiration. However, data scarcity is very often an issue in catchment modelling, which is particularly the case when it comes to weather data other than precipitation and temperature. As a result, data parsimonious evaporation methods have also been appealing choices in many catchment modelling applications. Few other methods which also vary in data requirements include Makkink (1957), Priestley-Tailor (Priestley and Taylor, 1972), and Hargreaves (Hargreaves et al., 1985). The latter two methods are presented here: Priestley-Tailor equation (3.47) and Hargreaves equation (3.48), which are probably the most commonly recommended methods as alternatives to Penman and Penman-Monteith in most hydrological models (see Section 3.9)

$$L_v E = \alpha_{PT} \frac{s_e (R_N - G)}{s_e + \gamma} \tag{3.47}$$

$$E = 0.0023 R_a (T_{\max} - T_{\min})(T_{mn} + 17.8)^{0.5} \tag{3.48}$$

In the Priestley-Tailor equation (3.47), α_{PT} is a coefficient (dimensionless) and all other parameters are the same as in the Penman equation.

In the Hargreaves equation (3.48), R_a is the extraterrestrial radiation (both E and R_a are in same units as equivalent water evaporation), T_{max} and T_{min} are daily maximum and minimum temperature in °C and T_{mn} is taken as the average of T_{max} and T_{min} (see Hargreaves, 1994).

Clearly the Priestley-Tailor is a simplification to the Penman method as it excludes the vapour transfer term from the Penman equation and compensates that by introducing an empirical coefficient α_{PT}. The value of α_{PT} is greater than 1.0 in the absence of condensation. For potential evapotranspiration assuming a fully saturated surface, Priestley and Taylor (1972) recommended $\alpha_{PT}=1.26$.

The Hargreaves method takes advantage of temperature data which is more commonly available as a standard meteorological observation and for radiation energy it relies on the extraterrestrial radiation, whose daily local values can be estimated from the information of latitude of the location and date. This makes it probably the most popular temperature based potential evapotranspiration method used in hydrological modelling. The parameters of the Hargreaves equation may be considered for calibration in catchment modelling. There are also modified versions of the equation available (see, e.g. McMahon et al., 2013).

3.7 Actual evaporation methods

As discussed earlier, the general assumption to distinguish between potential and actual evaporation in hydrological modelling is that PET is controlled by climate and AET is controlled by climate (through PET) and availability of water. This assumption supports the use of PET as an upper limit to AET. Therefore, the simplest way to estimate AET is by comparing PET with the available water on the surface or soil moisture, say AW, at any given time period. If there is sufficient water to evaporate (including transpiration), i.e.

If AW $\geq$ PET, then AET is equal to PET.
If AW $<$ PET, then AET is equal to AW.

The lower limit to the AET is zero, that is when AW$=0$.

This is a lumped view of AET. So, the AET methods in hydrological modelling deal with how to calculate AET discriminately. In general, AET constitutes one or more of the following:

- Evaporation from open water surfaces (e.g. lakes or reservoirs, rivers and ponded water in surface depressions), E_{OW}
- Evaporation from canopy intercepted water, E_{CI}
- Direct evaporation of soil water at the surface, E_{Soil}
- Plant transpiration of soil water, E_{PT}
- Snow sublimation (in case of snow/ice covered surface), E_{Snow}

Surfaces fully covered with water can be separated from bare soil or soil with vegetation. So E_{OW} can be calculated separately, which is usually taken to be equal to the potential evaporation rate.

E_{CI}, E_{Soil} and E_{TP} are parts of the same area and evaporation from all three can take place at the same time. So, to estimate the total AET from soil surfaces, two common practices can be found. First is to apply an 'order' in which the evaporation is to be removed. Note that this is not what physically happens, but just a trick to make the implicit process explicit for the sake of computation. It is a common practice in most hydrological models to take the E_{CI} first and also treat it as open water evaporation. That means, if at any given time period, the storage of canopy intercepted water is CI, then

$$\text{If PET} > \text{CI, then } E_{CI} = \text{CI; otherwise, } E_{CI} = \text{PET.}$$

Second, for the residual PET (that is PET – E_{CI}), broadly two approaches can be distinguished. One is a lumped approach in which no distinction is made between direct soil water evaporation and plant transpiration. In this case, a total evapotranspiration from the soil, E_{Soil_tot} ($= E_{Soil} + E_{TP}$) is calculated applying a reduction factor (r_1) to the residual PET. Thus,

$$E_{Soil\_tot} = r_1(\text{PET} - E_{CI})$$

The reduction factor $r_1 \leq 1$ is usually based on the ratio of the current soil moisture to the maximum soil moisture capacity. If there are more than one soil layer, then either an order is specified (i.e. first from the top layer, then the second, and so forth), or it is shared between the layers using a certain criterion.

Another approach estimates E_{Soil} and E_{TP} separately by sharing the residual PET between them. It is hard to point out one standard approach for the partitioning, but in general it involves two steps. The first is to share the residual PET between them, which is to determine separate potential rates for soil water evaporation and plant transpiration. The second is to apply reduction factors to estimate the actual value for each (E_{Soil} and E_{TP}). One way to implement this approach is to divide the soil into a top layer for direct soil water evaporation, and a root zone for plant transpiration. Then divide the proportion of the residual PET for E_{Soil} and E_{TP} based on the vegetation density, usually using the leaf area index (LAI) or in combination with vegetation cover fraction as a basis. Higher the vegetation density, higher the proportion for E_{TP}. Depending on the vegetation type, the root zones can also vary. So, for potential transpiration it is more appropriate to consider more vegetation characteristics, such as root depth and density, vegetation cover fraction and LAI. It may also be estimated using the Penman-Monteith equation with an appropriate canopy resistance for the type of vegetation.

For the reduction factor for E_{Soil}, the same approach as in the lumped soil evaporation can be used, which is based on the current soil moisture level compared to the maximum allowable for that layer. For E_{TP} the reduction can be based on the status of the soil moisture in the root zone for plant uptake. In the case of snow-covered surface, E_{Soil} is replaced by E_{Snow}.

The approaches to AET described here are only few examples that are used in hydrological modelling. Approaches used in various models vary in details and complexity, which are indicated in Table 3.1 for the 16 models reviewed.

3.8 Interception methods

Canopy interception is the portion of precipitation that is intercepted by vegetation (particularly the leaves) before reaching the ground. Its significance in evaporation estimation is well known in hydrological modelling, particularly in forested and densely vegetated catchments. But how much rain water the vegetation may intercept is a difficult parameter to model and so is to predict if and what fraction of the intercepted rainfall may fall on the ground before it gets evaporated. Because of these uncertainties almost all hydrological models use a rather simple approach to model interception. Although the details of the approaches vary, the main idea is to specify a maximum interception depth ($\mathrm{CI}_{\mathrm{max}}$), commonly in mm, for the given type of vegetation, and apply a reduction factor to the maximum interception to estimate interception potential ($\mathrm{CI}_{\mathrm{pot}}$) for a given time step (usually a day). For the reduction factor, the ratio of the LAI of the given day to the maximum LAI is commonly used. Accordingly, for any given day (using daily time step) the dry canopy (meaning without initial canopy storage) interception potential is given by

$$\mathrm{CI}_{\mathrm{pot,day}} = \mathrm{CI}_{max}\frac{\mathrm{LAI}_{\mathrm{day}}}{\mathrm{LAI}_{\mathrm{max}}} \tag{3.49}$$

Note that if there is canopy storage remaining from the previous day (wet canopy), that needs to be deducted to estimate the potential for that day. Finally, canopy interception for the day is given by

$$\mathrm{CI}_{\mathrm{day}} = \mathrm{Min}\left(\mathrm{CI}_{\mathrm{pot,day}}, P_{\mathrm{day}}\right) \tag{3.50}$$

where P_{day} is the precipitation of the day. As also discussed in Section 3.7, evaporation is first taken from the canopy interception at the potential rate. Once the evaporation is taken, the remaining canopy storage (i.e. CI - E_{CI}), which is usually a very small quantity (or even zero), is either added to the throughfall or carried over as storage to the next time step or divided between the two.

Note that the interception method described here is rather simple and numerically explicit approach but common in many catchment hydrological models. More formal descriptions of canopy interception models can be found in Rutter et al. (1971), Gash et al. (1995) and Galdos et al. (2012).

3.9 How different catchment models treat evaporation?

Brief descriptions of methods and approaches used in 16 hydrological models for evaporation are presented here (Table 3.1). The selected models are among the widely used models, but the list is not exhaustive, and the purpose of presenting the table is not for giving the author's judgement about the models. It is simply intended to describe the key concepts of the methods and approaches used in the models. The descriptions attempt to capture the main essence or features, but may not be detailed enough to provide complete details of the methods, for which the referenced literature should be consulted.

Table 3.1 Evaporation methods used in various hydrological models.

Modelling software	Method available	Additional information
BROOK90	Calculates evaporation using the Shuttleworth-Wallace approach (Shuttleworth and Wallace, 1985), whereby the P-M equation is applied separately to estimate evaporation from the soil and transpiration from plants as well as from canopy intercepted precipitation using appropriate resistances for different surfaces (Federer, 2002; Federer et al., 2003). See also Dingman (2002). It also estimates evaporation separately for day and night time. It does not estimate PET.	
CASC2D	For continuous simulation, it computes evapotranspiration using the Penman-Monteith equation on hourly time steps (Ogden, 1998; Downer et al., 2002).	
CHARM (also WATFLOOD)	PET is based on the Priestley-Tailor or Hargreaves method. AET is estimated from the PET applying three reduction factors, each for moisture available in the soil, soil temperature and vegetation types (tall or short). Evaporation from canopy intercepted water is also considered such that the intercepted water, when present, is taken first before the vegetation transpiration (Kouwen, 2018).	CHARM: Canadian Hydrological and Routing Model
Flo2D	Includes evaporation from open water surfaces (e.g. major river systems), but that is based on user input of monthly evaporation values (estimated externally) (FLO-2D, 2003).	

Table 3.1 Evaporation methods used in various hydrological models—cont'd

Modelling software	Method available	Additional information
HBV	PET is user input data (to be estimated externally) and used as an upper limit to compute AET. When soil moisture is between a specified value (a model parameter called LP) and field capacity, AET=PET, and when it is less than LP, AET is estimated applying a liner reduction factor to the PET. There is no separate calculation of plant transpiration (Bergström, 1992; Seibert, 2005).	
HEC-HMS	Evaporation is estimated only with the continuous simulation modules. Several methods are available for PET including Penman-Monteith, Priestley-Tailor and Hargreaves, and externally estimated daily PET as a direct input is also possible. The PET from either method will be used as an upper limit to calculate AET from the canopy interception and/or soil water depending on the basin loss method used (Hydrologic Engineering Centre, 2021). See also HEC-HMS Technical Reference (Feldman, 2000) and release notes v.4.1 through v.4.4.0.	
LISFLOOD	PET is calculated for three categories: potential reference evapotranspiration for vegetated area, potential evaporation for open water and potential evaporation from bare soil. PET estimates, however, are direct input to the model, but can be computed using its preprocesser LISVAP, which has the Penman-Monteith and Hargreaves methods for PET estimation. The PET is used to calculate actual evaporation from different land cover types. Evaporation and transpiration from soil and evaporation from intercepted water are separately computed (Burek et al., 2013a, 2013b).	
MIKE SHE	PET is used as reference ET as input data to the model (either from observation station or externally computed) together with crop	

Continued

Table 3.1 Evaporation methods used in various hydrological models—cont'd

Modelling software	Method available	Additional information
	coefficients. The Penman-Monteith based FAO-56 reference ET method is recommended in the user manual. The AET is primarily based on Kristensen and Jensen (1975) methods, which are empirically derived relationships to compute AET from PET. In the Kristensen-Jensen method, the plant transpiration and evaporation from soil water are computed separately. The former is based on an empirical relationship with LAI, soil moisture, plant root distribution (with user defined parameters), and the reference ET, and the latter on coefficients related to the soil moisture status as well as already computed plant transpiration. ET estimates from snow (sublimation) and canopy intercepted water are separately calculated, both as a fraction of the reference ET. A simpler version of AET estimation (as fractions of the reference ET) is also available in the simpler 2-layer subsurface model (DHI, 2017a, 2017b). (See also https://manuals.mikepoweredbydhi.help/2017/MIKE_SHE.htm.)	
NAM	PET is a meteorological input to the model and used as an upper limit to the AET. AET is first taken directly from the available water in the surface storage. If the PET is not met from the surface storage, the remaining PET demand is taken from the root zone water with a reduction factor defined by the ratio of water storage in the root zone to the maximum root zone storage (DHI, 2017c). Note that the description presented here is based on the NAM model used in MIKE 11, the river modelling software of DHI (https://manuals.mikepoweredbydhi.help/2017/MIKE_11.htm)	
PCR-GLOBWB	Computes potential reference ET based on the Penman-Monteith equation as in FAO-56 (Allen et al., 1998). The reference ET is partitioned into potential	This description is based on PCR-GLOBWB 1.0. Currently version 2.0

Table 3.1 Evaporation methods used in various hydrological models—cont'd

Modelling software	Method available	Additional information
	evaporation from soil water and potential transpiration. Reductions to both are applied to compute actual bare soil evaporation and transpiration based on the soil water storage status. Actual ET from external sources can also be direct input to the model (Van Beek and Bierkens, 2009).	is also available https://globalhydrology.nl/research/models/pcr-globwb-2-0/), but the key approach/methods for evaporation are the same (Sutanudjaja et al., 2018).
PRMS (Ver. 4) also used in GSFLOW	Several methods are available for PET including Penman-Monteith, Priestley-Tailor and Hargreaves-Samani (Hargreaves and Samani, 1982). AET is computed from various storages/surfaces with a specified order. Before the soil layer, evaporation is taken from intercepted rain/snow, water stored in impervious areas and surface depression storages. The soil layer from where evaporation and transpiration are extracted (called a "capillary reservoir") is divided in to two zones: the (upper) recharge zone and lower zone. To satisfy the remaining PET, soil evaporation is taken first from the recharge zone followed by transpiration from the lower layer. In both cases, the reduction factor is proportional to the ratio of the available soil water to maximum soil water in the respective zone (Markstrom et al., 2008, 2015). See also earlier versions in Leavesley et al. (1983, 1996).	PRMS: Precipitation Runoff Modelling System (Markstrom et al., 2015). GSFLOW: Coupled Ground-water and Surface-water Flow Model (Markstrom et al., 2008)
RRI model	Evapotranspiration is optional input to the model. If given, the same amount specified in the input file will be extracted from the surface/subsurface water storages (Sayama, 2017).	
SWAT	PET can be computed with three methods: Penman-Monteith, Priestley-Tailor and Hargreaves, or can be supplied directly as input data. Using the PET as an upper limit, AET is first taken from the canopy interception, and the remaining part from soil layers as soil evaporation and transpiration. The	

Continued

Table 3.1 Evaporation methods used in various hydrological models—cont'd

Modelling software	Method available	Additional information
	potential (or maximum) transpiration is estimated from the PET adjusted for evaporation from canopy water using a factor based on the LAI. Soil evaporation can take place from different layers. An exponential type function is used to determine the percentage of the total soil evaporation demand to be met from soil layers at different depths. If an area is under snow, sublimation also contributes to the total evaporation (Neitsch et al., 2011).	
VIC model	PET is based on the Penman-Monteith equation. AET constitutes evaporation from canopy intercepted water, transpiration and evaporation from bare soil. The canopy evaporation and transpiration are largely controlled by LAI and aerodynamic resistance and resistances specific to the canopy. Bare soil evaporation is taken from the top soil layer only and controlled by infiltration capacity and saturated fraction of the soil. The PET essentially serves as the upper limit to the three AET components (Liang et al., 1994; Gao et al., 2010; Lohmann et al., 1998). See also the VIC version 5 webpage: https://vic.readthedocs.io/en/master/	VIC (Variable Infiltration Capacity) home page: https://vic.readthedocs.io/en/master/
WaSiM	PET can be computed from the Penman-Monteith method, but a few other simpler methods are also available. The PET is partitioned in to potential interception evaporation, potential transpiration and potential soil evaporation. In the TOPMODEL-approach version, a reduction to the soil evaporation is controlled by the soil water status, and in the Richards-equation version by the soil water content as well as capillary pressure (Schulla, 2021). Snow sublimation is also calculated separately as a part of the energy balance computation of snow melt/sublimation.	See also the version updates on the WaSiM webpage: http://www.wasim.ch/en/the_model/dev_details.htm

Table 3.1 Evaporation methods used in various hydrological models—cont'd

Modelling software	Method available	Additional information
Xinanjiang Model	PET is computed from pan evaporation input values using a reduction coefficient. The PET demand is first supplied from the upper soil layer as much as possible, and the remaining part from the second and third layers, respectively. In case of the evaporation extraction from the second and third layers, a reduction factor is applied based on the soil moisture status of the respective layer. There is no separated estimation of evaporation and transpiration (Zhao, 1992; Yao et al., 2012). Note that several studies have reported the application of alternative evapotranspiration methods with the Xinanjiang model (Fang et al., 2017).	

References

Allen, R.G., Pereira, L.S., Raes, D., Smith, M., 1998. Crop Evapotranspiration: Guidelines for Computing Crop Water Requirements. FAO Irrigation and Drainage Paper No. 56, FAO, Rome, Italy.

Bergström, S., 1992. The HBV model—its structure and applications. SMHI RH No 4. Norrköping.

Bowen, I.S., 1926. The ratio of heat losses by conduction and by evaporation from any water surface. Phys. Rev. 27, 779–787.

Brutsaert, W., 1982. Evaporation into the Atmosphere: Theory, History and Applications. D. Reidel Publishing Co., Dordrecht.

Brutsaert, W., 2005. Hydrology, A Introduction. Cambridge University Press, New York.

Burek, P., Van der Knijff, J., De Roo, A., 2013a. LISFLOOD distributed water balance and flood simulation model. revised user manual. JRC Technical Report 'EUR 26162 EN'.

Burek, P., Van der Knijff, J., De Roo, A., 2013b. LISVAP Evaporation Pre-processor for LISFLOOD water balance and flood simulation model. revised user manual. JRC Technical Report 'EUR 26167 EN'.

Chin, D.A., 2011. Thermodynamic consistency of potential evapotranspiration estimates in Florida. Hydrol. Process. 25, 288–301. https:/doi.org/10.1002/hyp.7851.

Chow, V.T., Maidment, D.R., Mays, L.W., 1988. Applied Hydrology, International Edition. McGraw-Hill, Singapore.

Cuffey, K.M., Paterson, W.S.B., 2010. The Physics of Glaciers, fourth ed. Elsevier.

Cutnell, J.D., Johnson, K.W., 2013. Introduction to Physics, ninth ed. John Wiley and Sons, Inc., Singapore.

De Bruin, H.A.R., 1982. The Energy Balance of the Earth's Surface a Practical Approach. PhD Thesis, University of Wageningen, The Netherlands.

DHI (2017a). MIKE SHE Volume 2: Reference Guide. DHI, Denmark. https:/manuals.mikepoweredbydhi.help/2017/Water_Resources/MIKE_SHE_Printed_V2.pdf; Accessed on 31 July 2021.

DHI (2017b). MIKE SHE Volume 1: User Guide. DHI, Denmark. https:/manuals.mikepoweredbydhi.help/2017/Water_Resources/MIKE_SHE_Printed_V1.pdf; Accessed on 31 July 2021.

DHI (2017c). MIKE 11 A modelling system for rivers and channels, Reference Manual. DHI, Denmark. https:/manuals.mikepoweredbydhi.help/2017/Water_Resources/Mike_11_ref.pdf; Accessed on 31 July 2021.

Dingman, S.L., 2002. Physical Hydrology, second ed. Prentice Hall, New Jersey.

Doorenbos, J., Pruitt, W.O., 1977. Guidelines for Predicting Cropwater Requirements. Irrig. and Drain. Paper 24 (Revised), Food and Agricultural Organizations of the United Nations, Rome.

Downer, C.W., Ogden, F.L., Martin, W.D., Harmon, R.S., 2002. Theory, development, and applicability of the surface water hydrologic model CASC2D. Hydrol. Process. 16, 255–275.

Fang, Y.-H., Zhang, X., Corbari, C., Mancini, M., Niu, G.-Y., Zeng, W., 2017. Improving the Xin'anjiang hydrological model based on mass–energy balance. Hydrol. Earth Syst. Sci. 21, 3359–3375. https:/doi.org/10.5194/hess-21-3359-2017.

Federer, C.A., 2002. BROOK 90: A simulation model for evaporation, soil water, and streamflow. http:/www.ecoshift.net/brook/brook90.htm.

Federer, C.A., Vörösmarty, C., Fekete, B., 2003. Sensitivity of annual evaporation to soil and root properties in two models of contrasting complexity. J. Hydrometeorol. 4, 1276–1290.

Feldman, A.D. (Ed.), 2000. Hydrologic Modelling System HEC-HMS Technical Reference Manual. US Army Corps of Engineers, Hydrologic Engineering Centre, Washington, DC.

FLO-2D, 2003. FLO-2D User Manual. Nutrioso.

Galdos, F.V., Alvarez, C., Garcia, A., Revilla, J.A., 2012. Estimated distributed rainfall interception using a simple conceptual model and moderate resolution imaging spectroradiometer (MODIS). J. Hydrol. 468, 213–228. https:/doi.org/10.1016/j.jhydrol.2012.08.043.

Gao, H., Tang, Q., Shi, X., Zhu, C., Bohn, T.J., Su, F., Sheffield, J., Pan, M., Lettenmaier, D.P., Wood, E.F., 2010. Water budget record from variable infiltration capacity (VIC) model. In: Algorithm Theoretical Basis Document for Terrestrial Water Cycle Data Records. UNSPECIFIED. Available for download from https:/eprints.lancs.ac.uk/id/eprint/89407.

Gash, J.H.C., Lloyd, C.R., Lachaud, G., 1995. Estimating sparse forest rainfall interception with an analytical model. J. Hydrol. 170 (1–4), 79–86. https:/doi.org/10.1016/0022-1694(95)02697-n.

Hargreaves, G.H., 1994. Defining and using reference Evapotranspiration. J. Irrigat. Drainage Eng. 120 (6).

Hargreaves, G.H., Samani, Z.A., 1982. Estimating potential evapotranspiration. J. Irrigat. Drainage Eng. 108 (3), 225–230. https:/doi.org/10.1061/JRCEA4.0001390.

Hargreaves, G.L., Hargreaves, G.H., Riley, J.P., 1985. Irrigation water requirements for Senegal River Basin. J. Irrigat. Drain. Eng. 111 (3), 265–275.

Hydrologic Engineering Centre, 2021. Hydrological Modelling System: HEC-HMS User's Manual. U.S. Army Corps of Engineers, Davis, CA.

Kelliher, F.M., Leuning, R., Raupach, M.R., Schulze, E.-D., 1995. Maximum conductances for evaporation from global vegetation types. Agric. For. Meteorol. 73, 1–16.

Kent, M., 2013. Advanced Biology, second ed. Oxford University Press, p. 276.

Kouwen, N., 2018. WATFLOOD/CHARM Canadian Hydrological and Routing Model. University of Waterloo, Canada. http:/www.civil.uwaterloo.ca/watflood/index.htm.

Kristensen, K.J., Jensen, S.E., 1975. A model for estimating actual evapotranspiration from potential evapotranspiration. Nordic Hydrol. 6, 170–188.

Leavesley, G.H., Lichty, R.W., Troutman, B.M., and Saindon, L.G. (1983). Precipitation-runoff modeling system—User's manual: U.S. Geological Survey Water Resources Investigation Report 83-4238; https:/pubs.usgs.gov/wri/1983/4238/report.pdf; Last accessed on 1 August 2021.

Leavesley, G.H., Restrepo, P.J., Markstrom, S.L., Dixon, M.J., and Stannard, L.G. (1996). The Modular Modeling System (MMS)—User's manual: U.S. Geological Survey Open-File Report 96–151; https:/pubs.usgs.gov/of/1996/0151/report.pdf; last accessed on 1 August 2021.

Liang, X., Lettenmaier, D.P., Wood, E.F., Burges, S.J., 1994. A simple hydrologically based model of land surface water and energy fluxes for general circulation models. J. Geophys. Res. 99 (D7), 14415–14428.

Lohmann, D., Raschke, E., Nijssen, B., Lettenmaier, D.P., 1998. Regional scale hydrology: I. Formulation of the VIC-2L model coupled to a routing model. Hydrol. Sci. J. 43 (1), 131–141.

Lutgens, F.K., Tarbuck, E.J., 2001. The Atmosphere, eighth ed. Prentice-Hall, Inc., New Jersey.

Makkink, G.F., 1957. Testing the Penman formula by means of lysimeters. J. Inst. Water Eng. 11 (3), 277–288.

Markstrom, S.L., Niswonger, R.G., Regan, R.S., Prudic, D.E., Barlow, P.M., 2008. GSFLOW–Coupled Ground-water and Surface-water FLOW model based on the integration of the Precipitation-Runoff Modeling System (PRMS) and the Modular Ground-Water Flow Model (MODFLOW-2005). U.S. Geological Survey Techniques and Methods 6-D1, 240 p.

Markstrom, S.L., Regan, R.S., Hay, L.E., Viger, R.J., Webb, R.M.T., Payn, R.A., LaFontaine, J.H., 2015. PRMS-IV, the Precipitation-Runoff Modeling System, Version 4, Techniques and Methods 6–B7. U.S. Geological Survey, Reston, Virginia.

McMahon, T.A., Peel, M.C., Lowe, L., Srikanthan, R., McVicar, T.R., 2013. Estimating actual, potential, reference crop and pan evaporation using standard meteorological data: a pragmatic synthesis. Hydrol. Earth Syst. Sci. Supplementary materials: http:/www.hydrol-earth-syst-sci.net/17/1331/2013/hess-17-1331-2013-supplement.pdf.

Monteith, J.L., 1965. Evaporation and environment. Symp. Soc. Exp. Biol. 19, 205–234.

Monteith, J.L., Unsworth, M.H., 2013. Principles of Environmental Physics, fourth ed. Academic Press.

Murray, F.W., 1967. On the computation of saturation vapour pressure. J. Appl. Meteorol. 6, 203–204.

Neitsch, S.L., Arnold, J.G., Kiniry, J.R., Williams, J.R., 2011. Soil and Water Assessment Tool Theoretical Documentation Version 2009. Texas Water Resources Institute. Available electronically from https:/hdl.handle.net/1969.1/128050.

Ogden, F.L., 1998. A Brief Description of the Hydrologic Model CASC2D. Univ. Connecticut. rev 2001.

Oke, T.R., 1987. Boundary Layer Climate, second ed. Routledge.

Oki, T., Kanae, S., 2006. Global hydrological cycles and world water resources. Science 313, 1068–1072. https:/doi.org/10.1126/science.1128845.

Penman, H.L., 1948. Natural evaporation from open water, bare soil and grass. Proc. R. Soc. Lond. A Math. Phys. Sci. 193 (1032), 120–145.

Penman, H.L., 1956. Evaporation: an Introductory survey. NJLS – Wageningen J. Life Sci. 4 (1), 9–29. https:/doi.org/10.18174/njas.v4i1.17768.

Priestley, C.H.B., Taylor, R.J., 1972. On the assessment of surface heat flux and evaporation using large scale parameters. Mon. Weather Rev. 100, 81–92.

Rutter, A.J., Kershaw, K.A., Robins, P.C., Morton, A.J., 1971. A predictive model of rainfall interception in forests, 1. Derivation of the model from observations in a plantation of Corsican pine. Agric. Meteorol. 9, 367–384.

Sayama, T., 2017. Rainfall-Runoff Inundation (RRI) Model, Version 1.4.2. International Centre for Water Hazard and Risk Management (ICHARM) and Public Works Research Institute (PWRI), Japan.

Schulla, J., 2021. Model description WaSiM, Version 10.06.00. Hydrology Software Consulting J. Schulla, Zurich. http:/www.wasim.ch/downloads/doku/wasim/wasim_2021_en.pdf.

Seibert, J., 2005. HBV Light Version 2 User's Manual. Department of Physical Geography and Quaternary Geology, Stockholm University.

Shuttleworth, W.J., Wallace, J.S., 1985. Evaporation from sparse crops-an energy combination theory. Q. J. R. Meteorol. Soc. 111, 839–855. https:/doi.org/10.1002/qj.49711146910.

Streeter, V.L., Wylie, E.B., 1983. Fluid Mechanics, First SI Metric Edition. McGraw-Hill, Singapore.

Sutanudjaja, E.H., van Beek, R., Wanders, N., Wada, Y., Bosmans, J.H.C., Drost, N., van der Ent, R.J., de Graaf, I.E.M., Hoch, J.M., de Jong, K., Karssenberg, D., López López, P., Peßenteiner, S., Schmitz, O., Straatsma, M.W., Vannametee, E., Wisser, D., Bierkens, M.F.P., 2018. PCR-GLOBWB 2: a 5 arcmin global hydrological and water resources model. Geosci. Model Dev. 11, 2429–2453. https:/doi.org/10.5194/gmd-11-2429-2018.

Tetens, O., 1930. Uber einige meteorologische Begriffe. Zeitsch. Geophys. 6, 297–309.

Thom, A., Oliver, H., 1977. On Penman's equation for estimating regional evapotranspiration. Q. J. Roy. Meteorol. Soc. 103, 345–357.

Thornthwaite, C.W., 1948. An approach toward a rational classification of climate. Geogr. Rev. 38 (1), 55–94. https:/www.jstor.org/stable/210739.

Trambauer, P., Dutra, E., Maskey, S., Werner, M., Pappenberger, F., van Beek, L.P.H., Uhlenbrook, S., 2014a. Comparison of different evaporation estimates over the African continent. Hydrol. Earth Syst. Sci. 18, 193–212. https:/doi.org/10.5194/hess-18-193-2014.

Trambauer, P., Maskey, S., Werner, M., Pappenberger, F., van Beek, L.P.H., Uhlenbrook, S., 2014b. Identification and simulation of space–time variability of past hydrological drought events in the Limpopo River basin, southern Africa. Hydrol. Earth Syst. Sci. 18, 2925–2942. https:/doi.org/10.5194/hess-18-2925-2014.

Van Beek, L.P.H., Bierkens, M.F.P., 2009. The Global Hydrological Model PCR-GLOBWB: Conceptualization, Parameterization and Verification, Report. Department of Physical Geography, Utrecht University, Utrecht, The Netherlands. http:/vanbeek.geo.uu.nl/suppinfo/vanbeekbierkens2009.pdf.

Wallace, J.M., Peter, V.H., 2006. Atmospheric Science: An Introductory Survey, 2nd. Elsevier Academic Press, Amsterdam.

Ward, R.C., Robinson, M., 2000. Principles of Hydrology, fourth ed. McGraw-Hill, London.

Wild, M., Folini, D., Schär, C., et al., 2013. The global energy balance from a surface perspective. Climate Dynam. 40, 3107–3134. https:/doi.org/10.1007/s00382-012-1569-8.

Yao, C., Li, Z., Yu, Z., Zhang, K., 2012. A priori parameter estimates for a distributed, grid-based Xinanjiang model using geographically based information. J. Hydrol. 468–469 (2012), 47–62. https:/doi.org/10.1016/j.jhydrol.2012.08.025.

Zhao, R.-J., 1992. The Xinanjiang model applied in China. J. Hydrol. 135, 371–381.

Zomer, R.J., Trabucco, A., Bossio, D.A., van Straaten, O., Verchot, L.V., 2008. Climate change mitigation: a spatial analysis of global land suitability for clean development mechanism afforestation and reforestation. Agric. Ecosyst. Environ. 126, 67–80.

CHAPTER

Models of unsaturated (vadose) zone

4

4.1 Role of the unsaturated (vadose) zone

When rain water falls on the land surface, some or most of it infiltrates into the soil. If the surface is a vegetated surface, the canopy intercepts part of the rainfall before it reaches the ground. The interception process is discussed in Chapter 3. The rain water that is not infiltrated either remains as temporary storage in surface depressions, or becomes available for surface runoff. The transport of the surface runoff from the catchment to river is discussed in Chapter 5. The rainfall that is infiltrated becomes part of the subsurface water. The subsurface is divided into an unsaturated (or vadose) zone and its underneath a saturated zone or groundwater. As we know a soil mass consists of solid soil particles and pores or voids, which are filled with water and air. The ratio of the volume of voids (V_v) to the total volume of a soil sample (V) is called porosity (n), that is $n = V_v / V$ (m^3/m^3). Porosity is generally higher in uniformly graded soil than in well graded soil, and higher in unconsolidated soils than in consolidated soil (Bouwer, 1978). To give an example, porosity values listed in Kruseman and De Ridder (2000) for unconsolidated soils range from 0.4–0.7 for clay, 0.35–0.5 for silt and 0.25–0.5 for sand. A saturated soil refers to the soil when all of its voids are filled with water. In an unsaturated soil the voids are partly filled with water and partly with air. Of course, if the soil is completely dry, all the pores are filled by only air. The amount of water present in the soil is conveniently expressed as the volumetric water content, θ, given by the ratio of the volume of water (V_w) in the soil to the total volume, that is $\theta = V_w / V$ (m^3/m^3). The range of variation of θ is from zero for completely dry soil and less than or at most equal to its porosity (in fully saturated soil). The ratio of the volumetric water content to its porosity represents the degree of saturation, usually expressed in percentage, that is $\theta/n \times 100 = V_w/Vv \times 100$. More on soil properties can be found in Bouwer (1978) and Nonner (2015).

The unsaturated zone is the soil layer from the surface all the way to the layer which is fully saturated (that is the saturated zone or groundwater). Although it is conveniently called an unsaturated zone, in reality it alters between partly to fully saturation conditions. For example, during long or intense rainfall, the top soil layer can be temporarily saturated. There exists also a thin layer above the saturated zone, called a capillary fringe, which remains in saturation due to capillary rise of groundwater (see also Chapter 6). Another name for this layer from the soil surface to the saturated groundwater is 'vadose zone', and it may also be a preferred name because

Catchment Hydrological Modelling. https://doi.org/10.1016/B978-0-12-818337-3.00007-6

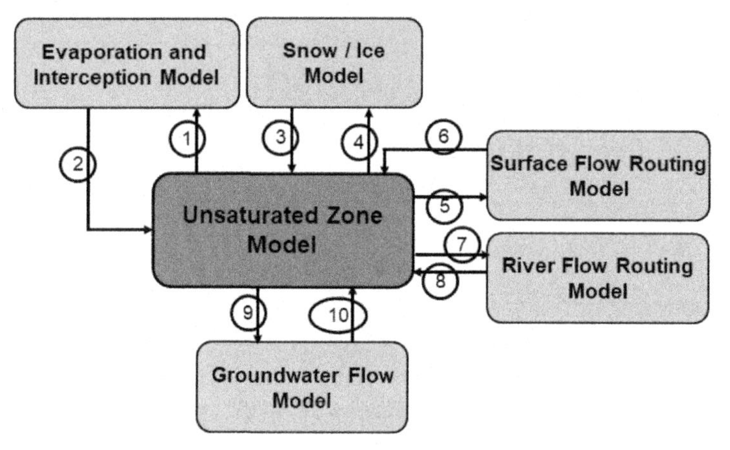

FIG. 4.1

Connection of the UZ Model to other components of the catchment model. Connections indicated by the numbers represent: (1) Water extraction from the soil by direct evaporation or plant transpiration, (2) Part of the intercepted precipitation (rain or snow) which falls to the ground surface, (3) Meltwater input, (4) Heat flux, (5) Surface flow input to the Surface Flow Routing, (6) *Re*-infiltration of the surface flow before reaching the streams, (7) Interflow (Q_{if}) input to the River Flow Routing, (8) Infiltration from river water, (9) Recharge input to the Groundwater Flow Model, (10) Moisture transport through capillary rise. Note that not all hydrological models represent all the connections shown here.

the entire layer is not always unsaturated (see Bouwer, 1978). In this book we mostly use unsaturated zone (UZ) or sometimes both. The saturated (or groundwater) flow is dealt with in Chapter 6.

The unsaturated zone is, in a way, the most important component of a hydrological model. It supports three major functions: (1) to partition the rain water into soil infiltration and surface flow (also called overland flow, or direct runoff), (2) to supply recharge (through percolation) to the groundwater (saturated zone) storage, and (3) to make water available for direct soil evaporation and plant uptake for transpiration. Additionally, although the water movement in the unsaturated zone is modelled as primarily a vertical flow process, some water may find its way to the streams, bypassing the vertical infiltration, through the process called interflow (see Section 4.3). As shown in Fig. 4.1, the UZ component is connected to every other component of a catchment hydrological model, which also explains the significance of this component in the model.

4.2 Unsaturated (vadose) zone flow methods

As we discussed earlier, the UZ is a central component in a catchment hydrological model. Essentially, it has connections to every other component in the model, but how the connection is actually represented depends on the type of the method used in the model. Broadly, three different approaches are used to model this component: empirical, conceptual and process based or physically based. Four methods are described here: the empirical Soil Conservation Services (SCS) Curve Number (SCS, 1985, see also in NRCS, 1986, 2004), conceptual Soil Water (Moisture) Accounting (SWA) (Burnash et al., 1973, Leavesley et al., 1983, see more in Section 4.2.2), physically based Green and Ampt (Green and Ampt, 1911, see also in Mein and Larson, 1973), and distributed physically based Richards equation (after Richards, 1931, see also in Smith and Woolhiser, 1971 and Mein and Larson, 1971). The SCS CN is probably the most commonly used empirical method in practice and also used in popularly used catchment models like Soil and Water Assessment Tool (SWAT) (Neitsch et al., 2011) and HEC-HMS (USACE, 2000) (see Table 4.1). The SWA, or more commonly SMA, is more like an approach than a particular method, in the sense that variation exists as to how the soil water accounting concept is implemented in a particular model. The Green-Ampt and Richards equations model the infiltration process in the UZ with physically derived equations, but the former use a more simplified representation.

4.2.1 SCS curve number

The SCS Curve Number (CN) method empirically relates the 'amount of direct runoff volume' from given rainfall to 'soil and land cover characteristics' and 'soil moisture condition'. The portion of the rainfall (from a rainfall event) estimated for direct runoff is also called precipitation excess or rainfall excess (R_e). In this method, a catchment area is assigned a number (the CN), whose value is determined as a function of soil, land cover and existing soil moisture condition, such that for the same amount of rainfall, different CN values result in different rainfall excesses (or direct runoff volumes). Higher the CN value, higher the direct runoff for the same amount of rainfall. The soil moisture condition, initially referred to as 'antecedent moisture condition' is more generally called 'antecedent runoff condition' (USACE, 1994). We will discuss about the upper and lower limits of CN values later in this section.

The basic principle assumed to derive R_e from R is that, once the runoff starts, the ratio of 'actual retention (S_a) to potential retention (S)' is equal to the ratio of 'excess rainfall (R_e) to total rainfall (R)'. Thus

$$\frac{S_a}{S} = \frac{R_e}{R} \tag{4.1}$$

where the units of all the parameters (S_a, S, R_e) are same as that of rainfall R, which is a cumulative volume over a given amount of time, but expressed in cumulative depth (length unit) over the area. With a given accumulation time period, R_e can also be expressed with the unit of depth per time [L T^{-1}], e.g., mm/h or mm/day. To maintain the volume conservation, the actual retention should be the difference between the total rainfall and the excess rainfall, i.e. $S_a = R - R_e$. This substitution yields

$$R_e = \frac{R^2}{R+S} \tag{4.2}$$

Note that the above relationship is applicable once the direct runoff is available from the rainfall. Any "losses" from a rainfall event before it is available for runoff is collectively referred to as initial abstraction (I_a), which includes interception in the canopy and surface depressions and infiltration (NRCS, 1986). To take this into account, R in Eq. (4.2) is replaced by $R - I_a$, which yields for $R > I_a$

$$R_e = \frac{(R - I_a)^2}{R - I_a + S} \quad (4.3)$$

If R is less than or equal to I_a, $R_e = 0$, so no surface runoff will generate. The initial abstraction is easy to describe but hard to accurately quantify. In most applications of the SCS CN method, it is approximated to 20% of S, i.e.

$$I_a = 0.2S \quad (4.4)$$

As we can see, CN is not used directly in the estimation of R_e (Eq. 4.3), but used in an intermediate step to estimate the potential retention, S. The conversion relationship between S and CN is reciprocal, given by

$$S = 25.4\left(\frac{1000}{\text{CN}} - 10\right) \quad (4.5)$$

Note that CN is a number with no unit, and in the above equation, the 25.4 is for the unit conversion of S from inch to millimetre, so here S is in mm. In the original equation, S is in inch given by CN = 1000/ (S – 10). From Eq. (4.5) we can see that the maximum value of CN (which is 100) results in zero retention, which means from Eq. (4.3), all the rainfall after initial abstract becomes available for surface runoff (i.e. $R_e = R - I_a$). When the initial abstraction is also zero, then $R_e = R$. The maximum limit for S is not defined, but CN = 0 makes S infinite, so strictly speaking, CN must be larger than 0. However, in practice, commonly used lower limit for CN is around 30.

In Fig. 4.2, R_e is plotted against R for different CN values based on Eqs. (4.3)–(4.5). As we can see, a CN value of 100 would generate 100 mm of accumulated excess rainfall (surface runoff volume) from the same amount of accumulated rainfall depth. Whereas, a CN value of 70, for example, would generate just about one-third of that (33 mm), and CN of 40 would generate just about 1.4 mm. This illustrates that CN is a very sensitive parameter for surface runoff, so care should be taken in model calibration for CN values. We will discuss about model calibration in Chapter 9.

A major difficulty in applying the CN method is to relate its value with the antecedent moisture or antecedent runoff condition (ARC) mentioned earlier. To take this into account SCS (1985) (see also NRCS, 1986) defines three types of CN values (CN I, CN II and CN III) for three ARC conditions (ARC I, ARC II and ARC III), respectively. ARC (II) represents an average condition of antecedent runoff, and ARC (I) and ARC (III) represent dry (low runoff potential) and wet (high runoff potential) conditions of antecedent runoff, respectively. Thus, CN (I) < CN (II) < CN (III). Tables of CN (II) values for different soil and land cover types and adjustments for ARC (I) and ARC (II) are available in USACE (1994).

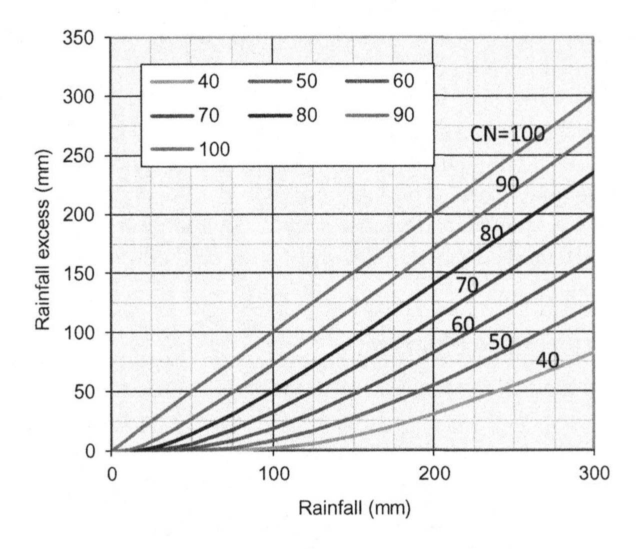

FIG. 4.2

Rainfall excess (R_e) verses rainfall depths (R) computed for different Curve Number values (CN=40, 50, 60, 70, 80, 90, and 100) using Eqs. (4.3) and (4.5). Initial abstraction, I_a=0.2S is used (Eq. 4.4).

The SCS CN method is basically intended for an event-based runoff estimation, and when it is applied as event-based, such as in HEC-HMS, the initial abstraction is considered as a "loss" from the rainfall. But the CN method is also used in continuous simulation model, such as in SWAT. In such a model (continuous simulation), the initial abstraction may constitute interceptions (canopy and surface depressions) and infiltration, which becomes part of the soil moisture.

4.2.2 Soil water (moisture) accounting

The SWA or SMA in hydrological modelling is often linked to the Sacramento Soil Moisture Accounting model (SAC-SMA) used by the NOAA National Weather Service (NWS) for operational river and flash flood forecasting (see II.3-SAC-SMA, 2002; Koren et al., 2007). However, the concept of accounting soil water in catchment modelling is similar to that started by Crawford and Linsley (1966) and Burnash et al. (1973). The "soil moisture accounting" terminology is also used in PRMS (Leavesley et al., 1983, see also Markstrom et al., 2015) and in HEC-HMS. Here the SWA is described as a more general approach used in various catchment models (see Table 4.1) to conceptually model unsaturated zone rather than a particular procedure used in the SAC-SMA model.

The key concept of the SWA approach is to divide the subsurface into a number of layers (also referred to as reservoirs), each with a specified water storage capacity, and infiltration capacity. Then to model the transport of water from one layer to another using a relationship (often nonlinear) based on the current water storage levels, storage capacities and infiltration (percolation) capacities of the two layers (or reservoirs) exchanging water. For example, the methods used in BROOK90 (Federer, 2002), HBV (Bergström, 1992), NAM (Nielsen and Hansen, 1973; DHI, 2017b) and Xinanjiang (Zhao, 1992) models (Table 4.1) may be considered as variations of the SWA approach. Because actual equations used to implement the approach vary from model to model, details of such methods are not presented here.

4.2.3 Green and Ampt equation

The Green and Ampt (also written as Green-Ampt) equation is one of the earliest but still commonly used methods to estimate infiltration as a function of time. It is derived based on the physically based Darcy's law to model the infiltration process in unsaturated soil. The Darcy's law states that the water flux (q) through a porous medium is proportional to the hydraulic gradient across the distance travelled, i.e. dh/dL, where h is the hydraulic head and L is the distance, or for the vertical soil column dh/dz. Thus, for the vertical soil column we can write

$$q = -K_s \frac{dh}{dz} \tag{4.6}$$

where K_s is the saturated hydraulic conductivity (dimension L/T), and as both h and z have the unit of length (e.g., m) q takes the same unit as K_s. In the above equation, the negative sign means that the flux q is taken positive upward. For infiltration rate (f), downward is taken positive, so substituting $q = -f$ in Eq. (4.6)

$$f = K_s \frac{dh}{dz} \tag{4.7}$$

The issue however is that infiltration is a highly complex process even in the case of fairly homogenous soil profile (Mein and Larson, 1973). As water moves down the soil column, an interface develops between the newly wet soil layer and the soil layer that is yet to be reached by the infiltrating water. This interface surface is called a "wetting front". However, the progression of the wetting front is non-uniform. Therefore, to apply the Darcy's law one assumption made in the Green-Ampt equation is a constant wetting front (see Fig. 4.3). The second assumption applied is that there is a continuous supply of water (or ponding) for infiltration at the surface.

Let us assume the elevation at the soil surface is z_0 and hydraulic head h_0, and at the wetting front the elevation and hydraulic head are z_{wf} and h_{wf}, respectively. This gives $dh = h_{wf} - h_0$ and $dz = z_{wf} - z_0$, and substituting in Eq. (4.7), we can write

$$f = K_s \left(\frac{h_{wf} - h_0}{z_{wf} - z_0} \right) \tag{4.8}$$

Assuming the elevation at the surface $z_0 = 0$, and z is negative downward, the length of the soil column from the surface to the wetting front $L_{wf} = -z_{wf}$. The hydraulic

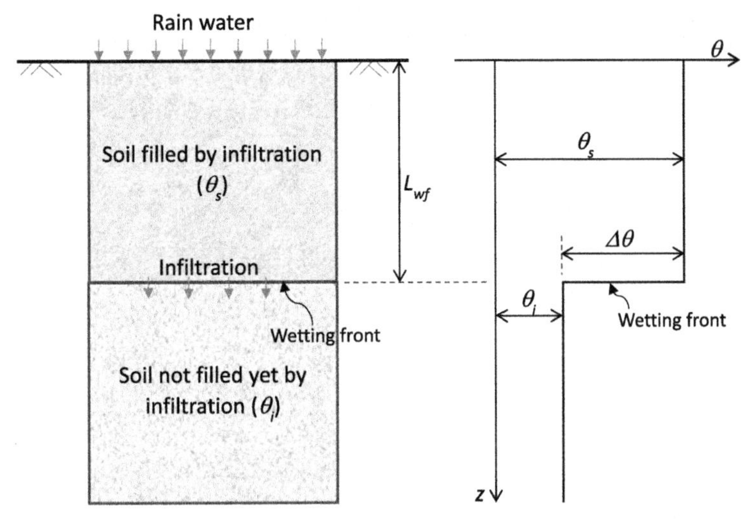

FIG. 4.3

Soil profile during infiltration (left) and idealized soil water content distribution for the Green and Ampt method (right). Sign conventions used are depth (z) downward negative and infiltration (f) downward positive.

head at the wetting front is negative and represented by the sum of the soil matric potential (ψ_{wf}) and the soil depth above it, which is L_{wf}. Thus, $h_{\mathrm{wf}} = -\psi_{\mathrm{wf}} - L_{wf}$. Furthermore, the ponding depth above soil surface is assumed negligible for the head at the surface, so $h_0 = 0$. Substituting these in Eq. (4.8), we obtain

$$f = K_s\left(\frac{\psi_{wf} + L_{wf}}{L_{wf}}\right) \tag{4.9}$$

In the above equations f is the infiltration capacity or infiltration rate (dimension L/T) at a particular time during infiltration. The cumulative amount of water that is infiltrated at a given time can be presented by F (t). Assuming that the soil column is fully saturated up to the depth of wetting front, the initial water content before the infiltration began is θ_i and water content at saturation is θ_s, we can write

$$F = L_{wf}(\theta_s - \theta_i) = L_{wf}\Delta\theta \tag{4.10}$$

Substituting L_{wf} in Eq. (4.9), the Green-Ampt equation for rainfall rate (r) greater than the infiltration potential rate (i.e. $r > f$) is given by

$$f = K_s\left(1 + \frac{\psi_{wf}\Delta\theta}{F}\right) \tag{4.11}$$

If the rainfall rate is less than the infiltration potential rate (i.e. $r \leq f_p$), the infiltration rate is assumed to equate to the rainfall rate, i.e. $f = r$.

4.2.3.1 Solution of the Green and Ampt equation

Note that in the Green and Ampt equation (Eq. 4.11), f represents the infiltration rate at a particular time, but it depends on the cumulative infiltration F until that time. So, to solve this relationship we first replace f by dF/dt, so that we will have only F in the equation, i.e.

$$\frac{dF}{dt} = f = K_s\left(1 + \frac{\psi_{wf}\Delta\theta}{F}\right) \tag{4.12}$$

By rearranging, we obtain

$$\frac{F}{\psi_{wf}\Delta\theta + F}dF = K_s dt \tag{4.13}$$

Finally, integrating Eq. (4.13) and substituting $F(t=0)=0$ yields

$$F(t) - \psi_{wf}\Delta\theta \ln\left(1 + \frac{F(t)}{\psi_{wf}\Delta\theta}\right) = K_s t \tag{4.14}$$

Note that this equation is an implicit function of $F(t)$ and needs to be solved iteratively using some kind of numerical technique (see, e.g., Chow et al., 1988). It has three parameters related to soil hydraulic properties: K_s, ψ_{wf}, and θ_s. With the cumulative infiltration $F(t)$ from Eq. (4.14), the infiltration rate f at time t can be estimated from Eq. (4.11) substituting $F = F(t)$.

In Eq. (4.10), $\Delta\theta = \theta_s - \theta_i$ is used. If the residual water content (θ_r) is considered, $\Delta\theta$ is computed with the effective water content (θ_e) and the effective saturation (s_e) given by

$$s_e = \frac{\theta_i - \theta_r}{\theta_s - \theta_r} \tag{4.15}$$

$$\theta_e = \theta_s - \theta_r \tag{4.16}$$

$$\Delta\theta = \theta_e(1 - s_e) \tag{4.17}$$

Estimates of Green-Ampt parameters have been presented by Rawls et al. (1983) for different soil types. More detailed discussion of hydraulic properties in soils and parameter relationships can be found in Brooks and Corey (1964).

The Green-Ampt equation also assumes that the soil above the wetting front is fully saturated, which is not the case when infiltration has just begun. In other words, it takes time before a layer of soil is fully saturated. To address this issue, Mein and Larson (1971, 1973) proposed a solution which has been used in some models, e.g. SWAT. It assumes that if $r < K_s$, all the rainwater will infiltrate without the surface being saturated. If $r > K_s$, the infiltration capacity is taken equal to the rainfall intensity, that is $f_p = r$. This leads to the relationship for cumulative infiltration before saturation (F_s) (from Eq. 4.12 with $f = r$ and $F = F_s$) as

$$F_s = \frac{\psi_{wf}\Delta\theta}{r/K_s - 1} \tag{4.18}$$

Then the time taken for the saturation (t_s) is given by

$$t_s = \frac{F_s}{r} \tag{4.19}$$

4.2.4 Richards equation

The Richards equation (after Richards, 1931) describes water movement in unsaturated soil combining the continuity equation with Darcy's law. It describes the variation of soil moisture over time and space (depth in case of one-dimensional vertical flow) and also allows the variation of hydraulic conductivity that varies with the variation of moisture content in the soil.

For the derivation of the equation, let us consider a controlled volume in a soil column. In three-dimensional representation, the continuity of moisture within the controlled volume can be described as following: the change in moisture content over time ($\partial\theta/\partial t$) must be balanced by the changes in fluxes (q's) across the three coordinate directions ($\partial q/\partial x$, $\partial q/\partial y$, $\partial q/\partial z$). Thus

$$\frac{\partial\theta}{\partial t} + \frac{\partial q_x}{\partial x} + \frac{\partial q_y}{\partial y} + \frac{\partial q_z}{\partial z} = 0 \tag{4.20}$$

where the fluxes q_x, q_y and q_z are in [L/T], and replacing them from the Darcy's Eq. (4.6) in the continuity Eq. (4.20), we obtain

$$\frac{\partial\theta}{\partial t} = \frac{\partial}{\partial x}\left(K(\theta)\frac{\partial h}{\partial x}\right) + \frac{\partial}{\partial y}\left(K(\theta)\frac{\partial h}{\partial y}\right) + \frac{\partial}{\partial z}\left(K(\theta)\frac{\partial h}{\partial z}\right) \tag{4.21}$$

where h is the total hydraulic head [L]. Considering only the vertical infiltration we exclude the terms $\partial/\partial x$ and $\partial/\partial y$ to yield

$$\frac{\partial\theta}{\partial t} = \frac{\partial}{\partial z}\left(K(\theta)\frac{\partial h}{\partial z}\right) \tag{4.22}$$

Replacing $h = \psi + z$, i.e. assuming the total hydraulic head is the sum of the soil matric potential and gravitational potential (with z negative downward and the reference level at the surface $z_0 = 0$), we obtain

$$\frac{\partial\theta}{\partial t} = \frac{\partial}{\partial z}\left(K(\theta)\frac{\partial\psi}{\partial z}\right) + \frac{\partial K(\theta)}{\partial z} \tag{4.23}$$

Note that if the moisture is taken up by plants for transpiration from the soil at a given depth, the plant extraction is a loss (or sink), W_s, in the moisture balance equation. Often the sink term comprises of three losses: evaporation from the top soil layer, plant water uptake (transpiration) from the root-zone (see also Chapter 3), and finally the recharge to the groundwater (saturated zone) which takes place at the bottom layer of the unsaturated zone. The plant transpiration loss can be from one or more layers depending on the size of the vertical computational grid and the depth of the plant roots. If the capillary rise at the top of the SZ is explicitly modelled as part of the

UZ, then the upward flux due to the capillary rise becomes a source term. With a sink term, the continuity Eq. (4.23) becomes

$$\frac{\partial \theta}{\partial t} = \frac{\partial}{\partial z}\left(K(\theta)\frac{\partial \psi}{\partial z}\right) + \frac{\partial K(\theta)}{\partial z} - W_s \tag{4.24}$$

This is a final form of the Richards equation for one-dimensional (vertical) infiltration. It has both θ and ψ as differential variable with respect to time. It can also be expressed with only $\partial\psi/\partial t$ or $\partial\theta/\partial t$, by defining a ratio, called the soil water capacity, $C = \partial\theta/\partial\psi$.

With $C = \partial\theta/\partial\psi$, we can write

$$\frac{\partial \theta}{\partial t} = \frac{\partial \theta}{\partial \psi}\frac{\partial \psi}{\partial t} = C\frac{\partial \psi}{\partial t} \tag{4.25}$$

Replacing this in Richards equation (4.24) gives

$$C\frac{\partial \psi}{\partial t} = \frac{\partial}{\partial z}\left(K(\theta)\frac{\partial \psi}{\partial z}\right) + \frac{\partial K(\theta)}{\partial z} - W_s \tag{4.26}$$

The $K(\theta)$ function is commonly expressed with respect to K_s and a relative hydraulic conductivity, $K_r(\theta)$ as $K(\theta) = K_s\, K_r(\theta)$. Substituting this in Eq. (4.26), we obtain

$$C\frac{\partial \psi}{\partial t} = K_s\frac{\partial}{\partial z}\left(K_r(\theta)\frac{\partial \psi}{\partial z}\right) + K_s\frac{\partial K_r(\theta)}{\partial z} - W_s \tag{4.27}$$

Note that K is a function of θ, and C is the slope of the ψ - θ curve. Saturated hydraulic conductivity (K_s) is often available for a given type of soil, but the actual hydraulic conductivity varies at different saturation levels for the same soil. The relationships ψ - θ and K - θ are nonlinear and that poses a significant challenge in the application of the Richards equation. Several empirical equations are available in the literature to approximate these relationships (Brooks and Corey, 1964; Campbell, 1974; Van Genuchten, 1980). One set of these equations based on Campbell (1974) are

$$K(\theta) = K_s K_r(\theta) = K_s\left(\frac{\theta}{\theta_s}\right)^{2b+3} \tag{4.28}$$

$$\psi = \psi_e\left(\frac{\theta_s}{\theta}\right)^{b} \tag{4.29}$$

$$K(\psi) = K_s K_r(\psi) = K_s\left(\frac{\psi_e}{\psi}\right)^{2+3/b} \tag{4.30}$$

where ψ_e is the air-entry potential, and b is an empirical parameter. In practice the parameters of these relationships maybe settled by calibration. If laboratory measurement data are available for ψ - θ and K - θ, they can be used as a lookup table with interpolation for values that are not in the table. More about these relationships and parameters can also be found in, e.g., Dingman (2002), Mawer et al. (2014), Montzka et al. (2017), and Guellouz et al. (2020).

Solving the Richards equation estimates ψ, θ and K for a vertical grid cell in the UZ at each time step, which then allows to estimate the infiltration rate and vertical flux from the grid cell applying Darcy's law.

4.2.4.1 Solution of the Richards equation

Implementation of the Richards equation with a catchment modelling is fairly completed. Lack of accurate soil hydraulic properties (usually non-homogenous soil) and initial and boundary condition data severely limits the success of the simulation. The finite difference numerical method is commonly applied for solving the equation with different numerical schemes and solution techniques. Two of such solution techniques in a good detail are available in Mein and Larson (1971) and Belmans et al. (1983). A slightly different solution scheme also in accessible details is available in DHI (2017a). The solution details presented in the SWAP (Soil-Water-Atmosphere-Plant) model reference manual (Kroes and Van Dam, 2003) is also comprehensive.

A brief summary of a finite difference solution is presented here. For a comprehensive detail, above mentioned references may be consulted. We start with the Richards Eq. (4.27), which has three partial differential terms. So, the first step is to replace them by respective finite difference terms. To write the finite difference equation we use notation j for the distance (depth) grid and n for the time grid (see Fig. 4.4). For example, ψ (j, n) represents the matric potential head in the soil column grid j at time level n. The notations $j-1$ and $j+1$ represent one grid point before (above) and after (below) j, respectively. Similarly, $n-1$ and $n+1$ are one time step (Δt) earlier and one time step later than n. If a variable is averaged between two time steps n and $n+1$, it is represented by $n+1/2$. Thus, the finite difference approximation equations for each of the differential terms of Eq. (4.27) can be written as

$$C\frac{\partial\psi}{\partial t} = C_j^{n+1/2}\frac{\psi_j^{n+1}-\psi_j^n}{\Delta t} \tag{4.31}$$

$$K_s\frac{\partial}{\partial z}\left(K_r\frac{\partial\psi}{\partial z}\right) = \frac{K_s}{\Delta z}\left(K_{r+}^n\frac{\left(\psi_{j+1}^{n+1}-\psi_j^{n+1}\right)}{\Delta z_+} - K_{r-}^n\frac{\left(\psi_j^{n+1}-\psi_{j-1}^{n+1}\right)}{\Delta z_-}\right) \tag{4.32}$$

$$K_s\frac{\partial K_r}{\partial z} = \frac{K_s}{\Delta z}\left(K_{r+}^n - K_{r-}^n\right) \tag{4.33}$$

In the above equations,

$$\left.\begin{aligned} \Delta z_- &= z_j - z_{j-1} \\ \Delta z_+ &= z_{j+1} - z_j \\ \Delta z &= \frac{\Delta z_-}{2} + \frac{\Delta z_+}{2} = \frac{z_{j+1}-z_{j-1}}{2} \end{aligned}\right\} \tag{4.34}$$

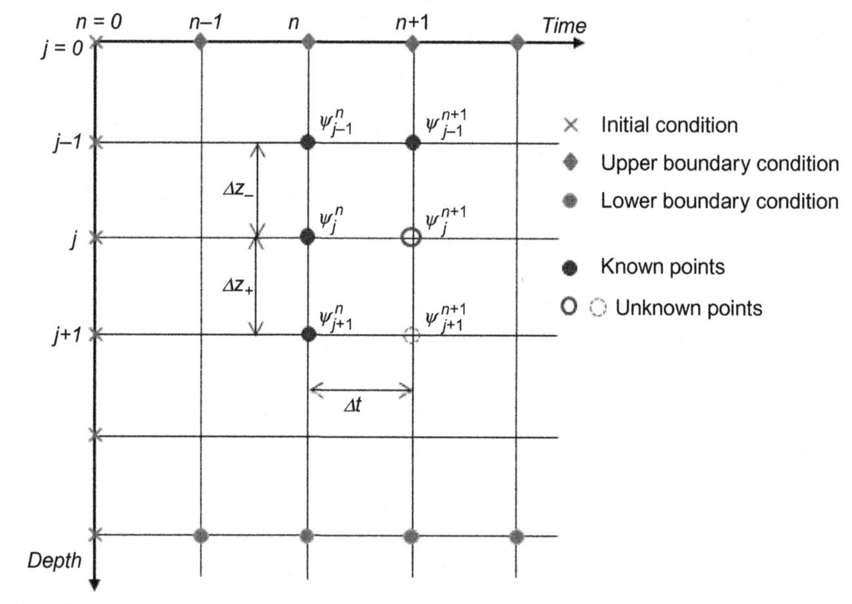

FIG. 4.4

Space (depth)–time discretization for a finite difference numerical solution of the Richards equation. Notations *n* and *j* represent time and space grids respectively. At the surface, $j=0$, and depth *z* is positive upward.

$$C_j^{n+1/2} = \frac{C_j^n + C_j^{n+1}}{2} \tag{4.35}$$

$$\left.\begin{aligned} K_{r-}^n &= \frac{K_{r,j-1}^n + K_{r,j}^n}{2} \\ K_{r+}^n &= \frac{K_{r,j}^n + K_{r,j+1}^n}{2} \end{aligned}\right\} \tag{4.36}$$

Note that K_{r-} and K_{r+} can be estimated in various ways. Different options are discussed in Mein and Larson (1971). In Eq. (4.36), space-centered values from time level *n* are used, that is, the arithmetic average of $(j-1, n)$ and (j, n) for K_{r-}, and (j, n) and $(j+1, n)$ for K_{r+} (see, e.g., Kroes and Van Dam, 2003). In MIKE-SHE they are calculated as the average of space-centered values from the *n* time step and running average of the values in each iteration (DHI, 2017a).

The space-time discretizations given above (Eqs. 4.31 and 4.32) use four points to approximate the differential variables, namely $(j-1, n+1)$, $(j, n+1)$, $(j+1, n+1)$ and (j, n), which is similar to that used in MIKE-SHE and SWAP models. In Mein and Larson (1971), six points are used to approximate the differential variables. Also, in the above equations all together (including the averaging of C, K_{r-} and K_{r+}) six grid points are used. This is an implicit numerical scheme (except for K_{r-} and K_{r+}) and its solution requires techniques involving iterations.

To give an example, for the first internal point, say (j_1, n_1), the three values containing time level $n=0$ come from the initial condition, that is ψ $(z, t=0)$, and the one with $(j-1, n+1)$ comes from the upper boundary condition at the surface, that is ψ $(z=0, t)$. There are still two unknowns with $(j, n+1)$ and $(j+1, n+1)$. The intention here is to calculate ψ $(j, n+1)$, so we still need to know an estimate for ψ $(j+1, n+1)$. At the lower boundary, which is at the top of the saturated layer, ψ $(z=L, t)$ (where L is the depth of the unsaturated zone) has to be defined. Therefore, ψ $(j+1, n+1)$ is known at the lower boundary. This means that we cannot solve this equation by one space-step at a time, but has to be solved simultaneously for all the space-steps from surface (top of the unsaturated layer) to the bottom of the unsaturated zone. For each grid point, the equation can be rearranged such that all the terms from time level $n+1$ are on the left-hand side (LHS) of the equation and all the rests on the right-hand side (RHS). These equations form a tridiagonal matrix, which can be solved efficiently (see, e.g., Remson et al., 1971; Harris and Stocker, 2006).

4.2.4.2 Initial and boundary conditions

Initial conditions required are ψ or θ for all the depth grids at $t=0$, that is ψ $(z, 0)$ or θ $(z, 0)$. Usually the initial water content is specified for all depths, and the corresponding matric potential (ψ) are estimated using the specified function, $\psi=f(\theta)$.

Two boundary conditions are required: upper boundary at the top soil layer and lower boundary at the top of the saturated zone (usually groundwater table). Rain water (including snowmelt) or already ponded water is the primary input for the upper boundary, and it results in two conditions. One when the top soil layer is yet to be saturated, and one when it is saturated and ponding has been created. These conditions can be specified, respectively, as a flux boundary (the infiltration rate at the top layer) and head boundary (ponding depth) at the top layer. The lower boundary condition at the top of the saturated layer is usually a head boundary based on the water table but it can also be a flux boundary (see, e.g., Ogden et al., 2015; Mathias et al., 2015).

4.3 Interflow methods

Interflow is the part of the subsurface unsaturated zone water that contributes to river flow before percolating into the saturated groundwater zone (except the perched saturated areas). In the literature this flow mechanism is referred to by various names, such as lateral flow, subsurface storm flow, throughflow, and secondary baseflow (Ward and Robinson, 2000; Weiler et al., 2006). Although definitions of the mechanism are to a certain extent ambiguous, Dingman (2002) has distinguished three types of mechanisms, namely (1) "sloping-slab" flow from perched saturated areas, (2) unsaturated Darcian flow in the soil matrix, and (3) pipe flow in macropores bypassing the unsaturated soil matrix. The role of the interflow is well recognized particularly in steep terrain and forested land cover in humid environment (see, e.g., Sloan et al., 1983; Weiler et al., 2006), but as the mechanisms vary, there is no one standard approach to model the interflow.

The surface flow contribution to the river flow is mostly limited to short periods in a year (e.g. when rain falls on saturated soil or rainfall intensity is larger than infiltration capacity), and it has faster response time. On the other hand, groundwater contribution is long lasting (e.g. in case of a perennial river, it is continuous throughout the year), but usually has slow response time. These two flow components (surface flow and groundwater flow) covers the top (peak flow) and bottom (baseflow) parts of a hydrograph. The interflow, usually with slower response time than surface flow but faster response time than groundwater baseflow, fills up the intermediate part of the hydrograph. Without this flow component with intermediate response time it is usually difficult to accurately simulate a catchment response to rainfall.

In some catchment models where subsurface is simulated as conceptual layers (reservoirs), runoff from one of the layers is usually referred to as interflow or lateral flow, e.g. in CHARM (Kouwen, 2018), HEC-HMS with the SMA method, NAM and Xinanjiang model (see Table 4.1). Some models, e.g. SWAT and PCR-GLOBWB (Van Beek and Bierkens, 2009), use a simplified method based on Sloan et al. (1983) (see also Sloan and Moore, 1984) for interflow.

4.4 How different catchment models treat unsaturated zone?

As we have seen there are various approaches and methods to model the unsaturated zone component. Methods and approaches used in 17 catchment models for the unsaturated zone are reviewed and briefly described here (Table 4.1). The selected models are among the widely used models, but the list is not exhaustive, and the purpose of presenting the table is not for giving author's judgement about the model. It is simply intended to describe the key concepts of the methods/approaches used in the models. The descriptions attempt to capture the main essence or features, but may not be detailed enough to provide complete details of the methods, for which the referenced literature should be consulted.

Table 4.1 Unsaturated zone (infiltration) methods in various catchment/hydrological models.

Modelling software	Method available	Additional information
BROOK90	The subsurface is divided into a number of soil layers, and the water available for infiltration is divided into the layers using a relationship $F_i = 1 - (z_i/Z)^b$ where z_i is the thickness of soil layer i, Z is the total soil thickness, b is a model parameter, F_i is infiltration through the bottom of layer i in fraction of the total	

Table 4.1 Unsaturated zone (infiltration) methods in various catchment/hydrological models—cont'd

Modelling software	Method available	Additional information
	water available for infiltration. Thus, the infiltration partition decreases exponentially with the depth from the surface. The infiltrating water that goes into layer *i* is given by $(z_i / Z)^b - (z_{i-1} / Z)^b$. Vertical flux of water through soil layers is modelled based on Darcy's law with a geometric mean of hydraulic conductivities of the adjacent layers (Federer, 2002; Federer et al., 2003)	
CASC2D	Two options are available: the Green and Ampt method and Green and Ampt with inter-storm redistribution of soil water (Ogden, 2001). See Ogden and Saghafian (1997) for the redistribution method.	
CHARM (also WATFLOOD)	Infiltration is based on the Philip (1954) formula which is similar to the Green-Ampt method with the ponded water (on the surface) added to the hydraulic head (Kouwen, 2018). Interflow is taken from the upper zone storage as a simple fraction of the moisture storage minus retention capacity of the upper zone times the topographic slope. The fraction value is a calibration parameter. The interflow and groundwater flow are estimated simultaneously.	CHARM: Canadian Hydrological and Routing Model
Flo2D	Uses the Green-Ampt method (FLO-2D, 2003)	
HBV	A type of soil water (moisture) accounting procedure with a soil water storage, and upper and lower groundwater reservoirs. The recharge to groundwater is determined from the rain water or snowmelt as a power function of the ratio of current soil moisture storage level to the maximum soil moisture storage (Bergström, 1992; Seibert, 2005). The power and the maximum soil moisture storage are model parameters to control the recharge. A third parameter, a lower soil moisture storage level, controls actual evaporation.	

Continued

Table 4.1 Unsaturated zone (infiltration) methods in various catchment/hydrological models—cont'd

Modelling software	Method available	Additional information
HEC-HMS	Several methods are available varying from simple parsimonious (e.g. initial abstraction and constant loss method) to popularly used Curve Number method and Green and Ampt method. It also has the conceptual Soil Moisture Accounting method for continuous simulation (USACE, 2000).	
LISFLOOD	The soil layer is divided into top soil and subsoil. The infiltration capacity is estimated from permeable land cover area using a procedure similar to that of Xinanjiang model. Water fluxes in the soil layer and out of the subsoil layer to groundwater are based on entirely gravity-driven flow assumption (Burek et al., 2013).	
MIKE SHE	It has the Richards equation and two simplified methods: Gravity Flow and Two Layer Water Balance (DHI, 2017a). In the gravity flow method, the pressure head term (primarily capillary force) is ignored and thus the flow in the unsaturated zone is assumed entirely due to gravity. The 2-Layer Water Balance method assumes two unsaturated zones using the formulation by Yan and Smith (1994).	
NAM	It has similar in concept to soil water (moisture) accounting with three layers: Surface Storage (SS), soil Root Zone (RZ) and groundwater (GW). Direct runoff and infiltration to the RZ are generated when SS exceeds the maximum storage limit. The proportion of direct runoff and infiltration is controlled by the moisture deficit in the RZ (DHI, 2017b). The moisture in the RZ are used for plant uptake, interflow as unsaturated zone runoff and input to GW storage. All of these are controlled by relative moisture storage in the RZ, that is, the ratio of the current moisture content to the maximum moisture content.	

Table 4.1 Unsaturated zone (infiltration) methods in various catchment/hydrological models—cont'd

Modelling software	Method available	Additional information
PCR-GLOBWB	It has two soil layers and underneath a groundwater storage layer. The first (top) soil layer controls surface runoff and infiltration to the second layer. The infiltration from layer 1 to 2 is calculated as unsaturated hydraulic conductivity of layer 1, and from layer 2 to groundwater storage (third layer) as unsaturated hydraulic conductivity of layer 2. Interflow is taken from the second soil layer with a simplified method based on Sloan and Moore (1984) (Van Beek and Bierkens, 2009).	
PRMS also used in GSFLOW	It divides the vertical layers into a surface layer (with pervious and impervious areas) and subsurface soil zone and underneath groundwater zone. The soil layer is divided into a preferential flow reservoir, capillary reservoir and gravity reservoir. Two types of surface flows are considered: infiltration excess (or Hortonian) flow from the soil layer and impervious surface and saturation excess from the soil layer. Fast and slow interflows are generated from the preferential reservoir and gravity reservoir, respectively. The gravity reservoir also drains to the groundwater layer (Markstrom et al., 2008, 2015).	PRMS: Precipitation Runoff Modelling System
RRI model	It uses the Green-Ampt equation. However, as RRI is primarily used for flood inundation modelling in rainfall events, infiltration is computed in limited conditions only, in particular in the areas with relatively small slopes. For example, in river valleys at early stages of rainfall events. (Sayama, 2017).	
SWAT	Two methods are available: the SCS Curve Number and Green and Ampt. The infiltrated water in the unsaturated soil layer can be taken up by plants. Recharge to the groundwater storage and lateral flow (to the river) are also taken from this layer. The lateral flow is computed based on the "kinematic storage model" by Sloan et al. (1983) (see also Sloan and Moore, 1984) (Neitsch et al., 2011).	

Continued

Table 4.1 Unsaturated zone (infiltration) methods in various catchment/hydrological models—cont'd

Modelling software	Method available	Additional information
UBC model	It has a simple conceptual model with moisture deficit and deficit decay function, which determine direct runoff and groundwater recharge (Quick and Pipes, 1977).	UBC: University of British Colombia
VIC	Surface runoff is based on the saturation excess of the soil layer, and vertical infiltration is based on the Richards equation in dispersion formulation (Liang et al., 1994; Gao et al., 2010). See also the VIC version 5 webpage: https://vic.readthedocs.io/en/master/	VIC (Variable Infiltration Capacity)
WaSiM	Unsaturated zone (soil storage) is modelled with two approaches: the TOPMODEL approach and Richards equation. In the former approach, the infiltration is estimated using a method based on Green and Ampt method (Schulla, 2021).	
Xinanjiang Model	It assumes three tension water storage (soil) layers: upper layer, lower layer and deepest layer. Runoff generation is similar to saturation excess runoff, i.e. direct runoff is generated when the storage capacity of the soil is filled to the field capacity level. The generated runoff is divided into three components (surface, interflow and groundwater) depending on the "free water storage" and free water storage capacity. A nonlinear curve is introduced to account for the variation of tension water storage within a catchment (Zhao, 1992; Yao et al., 2012; Fang et al., 2017).	

References

Belmans, C., Wesseling, J.G., Feddes, R.A., 1983. Simulation of the water balance of a cropped soil. SWATRE. J. Hydrol. 63, 271–286.

Bergström, S., 1992. The HBV model—its structure and applications. SMHI RH No 4. Norrköping.

Bouwer, H., 1978. Groundwater Hydrology. McGraw-Hill, New York.

Brooks, R.H., Corey, A.T., 1964. Hydraulic properties of porous media. Hydrology Paper No. 3, Colorado State University, Fort Collins, Colo.

Burek, P., Van der Knijff, J., De Roo, A., 2013. LISFLOOD distributed water balance and flood simulation model, revised user manual. JRC Technical Report 'EUR 26162 EN'.

Burnash, R.J.C., Ferral, R.L., McGuire, R.A., 1973. A Generalized Streamflow Simulation System—Conceptual Modeling for Digital Computers. U.S. Department of Commerce, National Weather Service and State of California, Department of Water Resources.

Campbell, G.S., 1974. A simple method for determining unsaturated conductivity from moisture retention data. Soil Sci. 117 (6), 311–314.

Chow, V.T., Maidment, D.R., Mays, L.W., 1988. Applied Hydrology. International Edition McGraw-Hill, Singapore.

Crawford, N.H., Linsley, R.K., 1966. Digital simulation in hydrology: Stanford Watershed Model IV. Technical Report No. 39, Department of Civil Engineering, Stanford University, Stanford, California.

DHI (2017a). MIKE SHE Volume 2: Reference Guide. DHI, Denmark. https://manuals.mikepoweredbydhi.help/2017/Water_Resources/MIKE_SHE_Printed_V2.pdf; Accessed on 31 July, 2021.

DHI (2017b). MIKE 11 A modelling system for rivers and channels, Reference Manual. DHI, Denmark. https://manuals.mikepoweredbydhi.help/2017/Water_Resources/Mike_11_ref.pdf; Accessed on 31 July 2021.

Dingman, S.L., 2002. Physical Hydrology, second ed. Prentice Hall, New Jersey.

Fang, Y.-H., Zhang, X., Corbari, C., Mancini, M., Niu, G.-Y., Zeng, W., 2017. Improving the Xin'anjiang hydrological model based on mass–energy balance. Hydrol. Earth Syst. Sci. 21, 3359–3375. https://doi.org/10.5194/hess-21-3359-2017.

Federer, C.A., 2002. BROOK 90: a simulation model for evaporation, soil water, and streamflow. http://www.ecoshift.net/brook/brook90.htm.

Federer, et al., 2003. Sensitivity of annual evaporation to soil and root properties in two models of contrasting complexity. J. Hydromereorol. 4, 1276–1290.

FLO-2D, 2003. FLO-2D User Manual. Nutrioso.

Gao, H., Tang, Q., Shi, X., Zhu, C., Bohn, T.J., Su, F., Sheffield, J., Pan, M., Lettenmaier, D.P., Wood, E.F., 2010. Water budget record from variable infiltration capacity (VIC) model. In: Algorithm Theoretical Basis Document for Terrestrial Water Cycle Data Records. UNSPECIFIED. Available for download from https://eprints.lancs.ac.uk/id/eprint/89407.

Green, W.H., Ampt, G.A., 1911. Studies on soil physics I. The flow of air and water through soils. J. Agric. Sci. 4, 1–24.

Guellouz, L., Askri, B., Jaffré, J., Bouhlila, R., 2020. Estimation of the soil hydraulic properties from field data by solving an inverse problem. Sci. Rep. 10, 9359. https://doi.org/10.1038/s41598-020-66282-5.

Harris, J.W., Stocker, H., 2006. Handbook of Mathematics and Computational Science. Springer, New York.

II.3-SAC-SMA (2002). Conceptualization of the Sacramento Soil Moisture Accounting Model. https://www.weather.gov/media/owp/oh/hrl/docs/23sacsma.pdf; Last accessed on 5 August, 2021.

Koren, V., Smith, M., Cui, Z., Cosgrove, B., 2007. Physically-based modifications to the Sacramento soil moisture accounting model: modeling the effects of frozen ground on the runoff generation process. NOAA Technical Report NWS 52. https://www.weather.gov/media/owp/oh/hrl/docs/NOAA_Technical_Report_NWS_52.pdf. (Accessed 6 January 2022).

Kouwen, N., 2018. Canadian Hydrological and Routing Model. User manual. Environment Canada. See also: http://www.civil.uwaterloo.ca/watflood/index.htm.

Kroes, J.G., Van Dam, J.C. (Eds.), 2003. Reference Manual SWAP Version 3.0.3. Altera -Report 773, Altera. Green World Research, Wageningen.

Kruseman, G.P., De Ridder, N.A., 2000. Analysis and Evaluation of Pumping Test Data, 2nd ed. International Institute for Land Reclamation and Improvement, Wageningen.

Leavesley, G.H., Lichty, R.W., Troutman, B.M., Saindon, L.G., 1983. Precipitation-Runoff Modeling System: User's Manual. Water-Resources Investigations Report 83-4238, Water Resources Division, Central Region, US Geological Survey, Denver, Colorado.

Liang, X., Lettenmaier, D.P., Wood, E.F., Burges, S.J., 1994. A simple hydrologically based model of land surface water and energy fluxes for general circulation models. J. Geophys. Res. 99 (D7), 14415–14428.

Markstrom, S.L., Niswonger, R.G., Regan, R.S., Prudic, D.E., Barlow, P.M., 2008. GSFLOW—Coupled Ground-water and Surface-water FLOW model based on the integration of the Precipitation-Runoff Modeling System (PRMS) and the Modular Ground-Water Flow Model (MODFLOW-2005). U.S. Geological Survey Techniques and Methods 6-D1.

Markstrom, S.L., Regan, R.S., Hay, L.E., Viger, R.J., Webb, R.M.T., Payn, R.A., LaFontaine, J.H., 2015. PRMS-IV, the Precipitation-Runoff Modeling System, Version 4, Techniques and Methods 6–B7. U.S. Geological Survey, Reston, Virginia.

Mathias, S.A., Skaggs, T.H., Quinn, S.A., Egan, S.N.C., Finch, L.E., Oldham, C.D., 2015. A soil moisture accounting-procedure with a Richards' equation-based soil texture-dependent parameterization. Water Resour. Res. 51, 506–523. https://doi.org/10.1002/2014WR016144.

Mawer, C., Knight, R., Kitanidis, P.K., 2014. Relating relative hydraulic and electrical conductivity in the unsaturated zone. Water Resour. Res. 51, 599–618. https://doi.org/10.1002/2014WR015658.

Mein, R.G., Larson, C.L., 1971. Modeling the Infiltration Component of the Rainfall-Runoff Process. Bulletin 43, Water Resources Research Centre, University of Minnesota.

Mein, R.G., Larson, C.L., 1973. Modeling infiltration during a steady rain. Water Resour. Res. 9 (2), 384–394.

Montzka, C., Herbst, M., Weihermüller, L., Verhoef, A., Vereecken, H., 2017. A global data set of soil hydraulic properties and sub-grid variability of soil water retention and hydraulic conductivity curves. Earth Syst. Sci. Data 9, 529–543. https://doi.org/10.5194/essd-9-529-2017.

Neitsch, S.L., Arnold, J.G., Kiniry, J.R., Williams, J.R., 2011. Soil and Water Assessment Tool Theoretical Documentation Version 2009. Texas Water Resources Institute. Available electronically from https://hdl.handle.net/1969.1/128050.

Nielsen, S.A., Hansen, E., 1973. Numerical simulation of the rainfall runoff process on a daily basis. Nord. Hydrol. 4 (3), 171–190. https://doi.org/10.2166/nh.1973.0013.

Nonner, J.C., 2015. Introduction to Hydrogeology, third ed. Taylor & Francis Group, London.

NRCS, 1986. Urban Hydrology for Small Watersheds, TR-55. Natural Resources Conservation Service, US Department of Agriculture.

NRCS, 2004. National Engineering Handbook, Part 630 Hydrology, Chapter 10. Natural Resources Conservation Service, US Department of Agriculture.

Ogden, F.L., 2001. A Brief Description of the Hydrologic Model CASC2D. Univ. Connecticut.

Ogden, F.L., Saghafian, B., 1997. Green and Ampt infiltration with redistribution. J. Irrigat. Drain. Eng. 123 (5), 386–393.

Ogden, F.L., Lai, W., Steinke, R.C., Zhu, J., Talbot, C.A., Wilson, J.L., 2015. A new general 1-D vadose zone flow solution method. Water Resour. Res. 51. https://doi.org/10.1002/2015WR017126.

Philip, J.R., 1954. An infiltration equation with physical significance. Soil Sci. 77 (2), 153–158.

Quick, M.C., Pipes, A., 1977. U.B.C. watershed model. Hydrol. Sci. J. 22 (1), 153–161.

Rawls, W.J., Brakensiek, D.L., Miller, N., 1983. Green-Ampt infiltration parameters from soils data. J. Hydraul. Eng. 109 (1), 62–70.

Remson, I., Hornberger, G.M., Molz, F.J., 1971. Numerical Methods in Subsurface Hydrology with an Introduction to the Finite Element Method. Wiley-International, New York.

Richards, L.A., 1931. Capillary conduction of liquids through porous mediums. Physics 1, 318–333.

Sayama, T., 2017. Rainfall-Runoff Inundation (RRI) Model, Version 1.4.2. International Centre for Water Hazard and Risk Management (ICHARM) and Public Works Research Institute (PWRI), Japan.

Schulla, J., 2021. Model Description WaSiM, Version 10.06.00. Hydrology Software Consulting J. Schulla, Zurich. http://www.wasim.ch/downloads/doku/wasim/wasim_2021_en.pdf.

Seibert, J., 2005. HBV Light Version 2 User's Manual. Department of Physical Geography and Quaternary Geology, Stockholm University.

Sloan, P.G., Moore, I.D., 1984. Modeling subsurface stormflow on steeply sloping forested watersheds. Water Resour. Res. 20, 1815–1822. https://doi.org/10.1029/WR020i012p01815.

Sloan, P.G., Moore, I.D., Coltharp, G.B., Eigel, J.D., 1983. Modeling Surface and Subsurface Stormflow on Steeply-Sloping Forested Watersheds. KWRRI Research Reports. 61 https://uknowledge.uky.edu/kwrri_reports/61.

Smith, R.E., Woolhiser, D.A., 1971. Mathematical Simulation of Infiltration Watersheds. Hydrology Paper, No. 47, Colorado State University.

Soil Conservation Service (SCS), 1985. National Engineering Handbook. Section 4-Hydrology, Washington, DC.

USACE, 1994. Flood-Runoff Analysis, EM 1110–2-1417. Office of Chief of Engineers, Washington, DC.

USACE, 2000. Hydrologic modelling system HEC-HMS, Technical Reference Manual. In: Feldman, A.D. (Ed.), US Army Corps of Engineers. Hydrologic Engineering Centre, Davis, CA.

Van Beek, L.P.H., Bierkens, M.F.P., 2009. The Global Hydrological Model PCR-GLOBWB: Conceptualization, Parameterization and Verification. Report, Department of Physical Geography, Utrecht University, Utrecht, The Netherlands http://vanbeek.geo.uu.nl/suppinfo/vanbeekbierkens2009.pdf.

Van Genuchten, M.T., 1980. A closed-form equation for predicting the hydraulic conductivity of unsaturated soils. Soil Sci. Soc. Am. J. 44, 892–898.

Ward, R.C., Robinson, M., 2000. Principles of Hydrology, fourth ed. McGraw-Hill, London.

Weiler, M., McDonnell, J.J., Tromp-van Meerveld, I., Uchida, T., 2006. Subsurface stormflow. In: Encyclopedia of Hydrological Sciences. Wiley Online Library, https://doi.org/10.1002/0470848944.hsa119.

Yan, J., Smith, K.R., 1994. Simulation of integrated surface water and groundwater systems—model formulation. JAWRA 30 (5), 879–890.

Yao, C., Li, Z., Yu, Z., Zhang, K., 2012. A priori parameter estimates for a distributed, grid-based Xinanjiang model using geographically based information. J. Hydrol. 468–469 (2012), 47–62. https://doi.org/10.1016/j.jhydrol.2012.08.025.

Zhao, R.-J., 1992. The Xinanjiang model applied in China. J. Hydrol. 135, 371–381.

CHAPTER

Models of surface (overland) flow routing 5

5.1 What is surface flow routing?

One of the several things the unsaturated zone model (Chapter 4) does is computes the volume of water that is available for direct runoff. The direct runoff is also commonly referred to as surface flow runoff, surface runoff, or overland flow. The rainfall water becomes available for direct runoff when it is in excess of interception (by plant canopy and surface depression) and infiltration. That is why it is also called rainfall excess (*R*e) or precipitation excess (*P*e). *R*e is used here instead of *P*e, because if the precipitation is in the form of snow, it needs to melt first before being available for runoff. It is expressed with the same unit as the rainfall rate, that is depth per unit time [LT^{-1}], usually mm/day or mm/h, generated over a certain area. So, to compute the water volume of surface runoff, say W_{sf}, we need to multiply *R*e by the area and time period (T). That is $W_{sf}=A\times Re \times T$, where A is the area [L^2] over which the rainfall excess *R*e is generated. This rainfall excess is an average water depth on a particular grid-cell (in case of a grid-based spatially distributed model), or on a catchment or subcatchment or even a certain area of the subcatchment (in case of a lumped or semidistributed model). The rainfall excess generated over an area of the catchment or the entire catchment has to travel a certain distance before it reaches the catchment outlet or drains to the river as surface runoff contribution to the river flow.

In the case of a distributed grid-based model, the *R*e travels from one grid-cell to one or more neighbouring cells, step by step, until it meets a river or a grid-cell marked as a river-cell. In case of a semidistributed model (catchments and subcatchments), it is in a way transferred directly to the inlet point of the river reach in that subcatchment or, if there is no river reach in that subcatchment, to the outlet point of the subcatchment. Once the surface runoff gets to the river, it will be transported down the river by the river routing model, which is discussed in Chapter 7. Now the question is how do we transport *R*e from one grid-cell to another, or from an entire subcatchment to its river reach or to the nearest river? The answer to this question is the model of surface flow routing. We need mathematical methods (or models) to transport the direct runoff volume from where (in the catchment) it is generated to the river or to the outlet of the catchment. These methods are discussed in Section 5.3.

As we can see in Fig. 5.1, the connection of the surface flow routing component to other components of a catchment model is rather simple. The surface flow volume is

Catchment Hydrological Modelling. https://doi.org/10.1016/B978-0-12-818337-3.00003-9

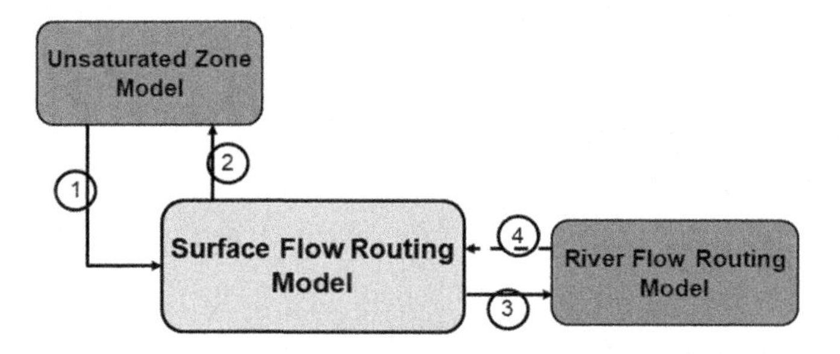

FIG. 5.1

Connection of the Surface Flow Routing (SFR) Model to other components of the catchment model. Connections indicated by the numbers represent: (1) *Re* input to the SFR, (2) Residual *Re* back to UZ, (3) Q_{sf} to the River Flow Routing, and (4) Overbank river flow to the floodplain area in the catchment. Connections between components other than SFR are not shown here.

computed in the unsaturated zone model, and transported to the river by the surface flow routing model. So it is connected to the unsaturated zone for its input and to the river flow routing model for its output. The input is *Re* and output is a surface flow rate that is added to the river discharge within a given time period (or computation time step). The flow rate that is added to the river discharge as surface flow is commonly represented by Q_{sf}, expressed in volume per unit time (e.g. m^3/s). Q_{sf} is one of the three major components of a river discharge. The other two components are interflow or lateral flow, Q_{if}, and base flow or groundwater flow, Q_{gw}. It may happen that not all of the *Re* generated in one time step gets to the nearest river during the same time step, and so a part of it can still evaporate, infiltrate or even freeze if the condition prevails. Which means that the connection between the unsaturated zone and surface flow routing can be two way, such that the residual *Re* (or the part of the *Re* not transported to the river cell) will be input to the unsaturated zone to add to the available water for surface evaporation, infiltration and freezing. Note however that not all catchment models use the two-way connection between the unsaturated zone and surface flow routing.

What about the connection with the river flow component, is it one-way or two-way? A two-way connection is shown in Fig. 5.1, but the connection from the river flow routing component to surface flow routing is indicated with a dotted line. When the river flow exceeds the bankfull discharge, water flows into the floodplain area in the catchment, but this process is handled only if a physically based (or process based) model is used for the surface flow routing.

5.2 Conceptualization of surface flow in a catchment

Conceptualization plays a key role in routing the surface flow in a catchment. The art side of modelling becomes as important as the science. We can see that the art and science go hand in hand. To understand why, imagine the last time you have seen surface (overland) flow happening, not in a river or a drainage channel, but anywhere else on an open land surface. It must have been during or just after a heavy rainfall. How deep was the flow? Was it all over the area you could see or only on some parts of the area? Could you see where was it beginning and where was it ending? Well, it could have been few centimetre or few inches depending on how intense and how long it was raining. Very likely it was covering only some parts of the area you could see. Where it was beginning could be hard to see, but you could probably see where it was ending. You might have seen a part of it draining to a nearby road drainage, to a small gully, or obstructed by a house and either changed its path or joined the drainage sewer of the house.

These details, e.g. the houses, road drainage, small gullies, etc. we usually do not see in a catchment model unless a very detailed one. So even in a spatially distributed model, we have to represent the effects of these details in an indirect way. To do that we typically ask ourselves what would have been the effects of those details in the flow routing. For example, the obstructions would slow down the flow and the gullies would speed it up. These subgrid level variabilities are treated as roughness factors: more obstructions (buildings, trees, large boulders, etc.) cause more resistance to flow, so higher roughness values; more gullies and drains may result in faster water flow, so lower roughness values. Note that some gullies may have depressions, which would actually create ponding until they are fully filled. The depressions exist not only in the gullies, but also elsewhere in the surface area. These are typically treated as surface depression storage, which means in the model the rainfall excess is routed from a grid-cell or catchment only when it exceeds the maximum possible depression storage in the grid-call/catchment. It is easy to say but way too difficult to actually implement these features in the model, and even if we do implement them, we usually lack data to realistically validate their effectiveness in model performance.

Knowing all these complexities, we may consider two ways to approach surface flow routing. One is the hard and laborious way, which is to build a detailed and fine resolution distributed model with two-dimensional (2D) fully dynamic flow routing equations. This is a possibility and there are models that allow to do this to a large extent, but would we have sufficient data to represent all these details? And how practical is it computationally? The answer is usually either 'not really' or 'yes but only for a small area' of not more that several square kilometres. Take one example on the dimensionality of the problem. If we have a catchment of $100\,km^2$ (which is usually a small catchment in most standards) and build a $100 \times 100\,m$ resolution model, we are talking about 10,000 grid-cells, which means computations on 10,000 cells every time step. To represent the type of variability we noted above, we are likely to realize that 100m resolution is actually not finer enough and may be tempted to go for a $10 \times 10\,m$ resolution if we have the topographic map or DEM of that detail. That means we end up having 1,000,000 computations per time step. Usually we run a hydrological model on a daily time step, but not if we have a surface flow routing with fully dynamic equations and 10m resolution grids. For such a model, we may be talking about time steps in the order of several minutes. Say, we run the model for 10 days of rainy season with a time step of 10 min, the model will have to carry out 1,440,000,000 computations.

Another way to approach surface flow routing is when we may be not interested in such details, but only interested to know how much of the rainfall excess generated in the catchment in one day gets to the nearest river the same day, irrespective of the route the flow takes. This basically means establishing a relationship with some kind of mathematical technique between the rainfall excess in a catchment and the amount of surface flow that the rainfall excess contributes to the river or to the catchment outlet in a particular time step or a day. Such a relationship is also called a transfer function; a function to transfer the rainfall excess in a catchment as surface runoff input to the river. One such approach has been in practice since the 1930s popularly known as the 'Unit Hydrograph' method, which we discuss in Section 5.3.1.

The physically based 2D flow concept may sound attractive but to apply it for surface flow on a natural catchment we have to make a lot of assumptions and simplifications. We are talking about the flow that occurs only a limited period of time, and often with a lot smaller depth compared to flow in a river channel. Unless the grid-cell is sufficiently small, the actual surface flow may not cover the entire cell area. Moreover, what about the slope from one cell to the next? What about the obstruction to the flow and the roughness due to trees, other vegetation, terraces that are constructed on the slope, etc.? Note also that hydrodynamic methods can be quite sensitive to sudden changes in slopes as well as channel geometry and roughness. Knowing all these, we can say that a realistic application of a detailed 2D surface flow routing is limited to a relatively small catchment area where high resolution DEM and land use data are available.

5.3 Surface flow routing or transfer methods

As discussed in Chapter 1, we can classify the surface flow routing methods as empirical, conceptual and process based or physically based methods. The unit hydrograph method (mentioned above) is a good example of an empirical method, which uses an empirical relationship—a transfer function—to transform the rainfall excess to surface runoff. There are different versions of UH or transfer function type methods available in the literature, see e.g. HEC-HMS technical document (Feldman, 2000). Note that in the grid-based distributed model, applying the unit hydrograph type method is not handy, because in principle the parameters of the relationships can be different for different grid-cells.

In the conceptual method, we attempt to recognize the major physical processes involved and represent them using usually simple (algebraic) but physically meaning equations. One example of conceptual surface flow routing is based on the simple estimation of the travel velocity of flow using the catchment topographic slope and surface resistance. This is often applied with a spatially distributed (grid based) model, in which the rainfall excess from a grid-cell is transported into one of the eight neighbouring cells with the lowest elevation. This method of flow travel route selection is popularly known as the 'steepest gradient' method. The actual transport to the cell with steepest gradient is based on some sort of velocity estimates, e.g. using a uniform flow type formula, such as Manning's formula. Sometimes a physically derived method is applied with some simplification and assumptions that are usually valid only on much smaller scale than the model is applied. For example, the application of kinematic wave momentum equation to spatially semidistributed or distributed model structure.

Some models use the kinematic wave based method for surface flow routing in a semidistributed model. The kinematic wave method is more commonly used for a river flood routing where there is clear boundary for the water flow. But when it is applied on the catchment level, it has to make an assumption of a channel which actually does not exist. So, in this case it is a "conceptual" application of a physically based kinematic wave method.

In the physically based approach the surface flow routing is treated as two-dimensional free surface flow and solved using 2D form of the St. Venant equations or shallow water equations. The physically based 2D method can only be applied with a grid-based distributed model with relatively small grid-size.

Between the lumped or semidistributed (with subcatchments) and distributed models, one major difference in the surface flow routing comes from the way the travel time is calculated. The travel time is related to the time lag between the time of rainfall excess generation and the surface runoff reaching the river. The rainfall excess available at the far upstream part of the (sub)catchment takes more time than the one near the outlet. Whereas in the distributed model the *R*e is routed from the cell where it is generated all the way down the outlet (cell by cell), in the lumped or semi-distributed model it is transferred based on one average velocity or lag-time for all the *R*e generated in different parts of the subcatchment.

5.3.1 Unit hydrograph and linear reservoir methods

5.3.1.1 Unit hydrograph

The key question in the UH type method (see, e.g. Feldman, 2000) is if there is a unit depth of *R*e (rainfall excess), what will be its response to the river discharge? There are two assumptions about the *R*e we need to understand. One, the unit depth of *R*e is the average depth over the entire catchment, subcatchment or a grid-cell, depending on whether the model is lumped, semidistributed or distributed. Two, the unit depth *R*e is uniform over the entire time period, say t. We can say that both of them are rarely true in a real case unless the area of the (sub)catchment or grid is very small and the time period is very short. But these are the types of assumptions we often make in hydrological modelling.

Once we know the *R*e over a given time period, the starting point for UH is that the volume *Re* is the same as the total surface runoff, Q_{sf}, it will produce. Only their timing and time rates can be different. Then we have to ask following questions: When will the runoff in the river start to appear due to the *Re*, within the same time period or later? How many periods does it last? And if it lasts for more than one time period, how does it distribute over the time periods?

If the discharge appears on the same time period and lasts on the same period as the Re, it will be the most likable situation for the modeller. Because then for any time period when the Re is available it will produce the same amount of runoff over the same time period. Then we only need to convert the unit, which is usually from mm of Re over a time period to m^3/s for Q_{sf}. For example, if the Re is in mm/d then the Q_{sf} in m^3/s $= Re$ (mm/day) $\times A$ (km^2) $\times 10^3$ / (24 × 3600).

In general, this assumption works well as long as the computational time step is sufficiently larger than the time of concentration, t_c. The time of concentration refers to the longest travel time of the runoff generated somewhere in the catchment to reach the catchment outlet. It is not necessarily the most distant point in the catchment from the outlet but from where the travel time is the longest (USDA, 2010a). Note however that the time of concentration is not the only thing that matters here. Even if the time step is larger than the time of concentration, the rainfall rate over the time period (if the period is large, e.g. several hours or a day) is unlikely to be uniform over the time period. If the rainfall happens to have occurred towards the end of the time period, most part of the Re will probably reach the river in the next time step. Similarly, if the time step is short and time of concentration is large, the discharge from the Re may continue for more than one time step. So by knowing from start and end time periods of surface flow input to the river discharge due to a given Re, we are kind of half-way to get the unit hydrograph. Suppose for a given catchment a Re produces surface runoff discharge for four time periods, beginning from the same time step to three more time steps. If we represent the fractions of Re that are distributed into these four time periods (t_1,..,t_4) by UH ordinates U_1,..,U_4, we can write

$$Re_1 = Re_1\,(U_1 + U_2 + U_3 + U_4)$$

where Re_1 is the rainfall excess generated in time period t_1. Note here that the UH ordinates have the same unit as the rainfall excess, e.g. mm/day (dimension L T^{-1}), and the sum of the ordinate values equal to 1. Then the surface flow produced by the Re_1 on each of the four time periods are

$$Q_{sf(1)} = Re_1 U_1, \ldots, Q_{sf(4)} = Re_1 U_4.$$

To make it a bit more complicated, suppose we have rainfall excess generated in two time periods t_1 and t_2, represented by Re_1 and Re_2 as shown below

$$Re_1 = Re_1\,(U_1 + U_2 + U_3 + U_4)$$

$$Re_2 = Re_2\,(U_1 + U_2 + U_3 + U_4)$$

So, the contribution of Re_2 will be added to $Q_{sf(2)}$, and so forth if there are more Re's. An example with the rainfall excess in three consecutive time steps are shown in Table 5.1 and illustrated in Fig. 5.2 with rainfall excess of 2mm/d in (1) just one day, (2) two consecutive days and (3) three consecutive days. The UH lasting for four days with $U_1 = 0.5$, $U_2 = 0.3$, and $U_3 = U_4 = 0.1$ are assumed. Note that these values are arbitrarily taken to simply illustrate the concept.

Table 5.1 The UH method: Surface runoff from rainfall excess on more than one time period. In this example the UH has four ordinates.

Rainfall excess Time (t_1...t_n)	Q_{sf} from Re_1	Q_{sf} from Re_2	Q_{sf} from Re_3	$Q_{sf(t)}$
1	Re_1U_1			$Q_{sf(1)} = Re_1U_1$
2	Re_1U_2	Re_2U_1		$Q_{sf(2)} = Re_1U_2 + Re_2U_1$
3	Re_1U_3	Re_2U_2	Re_3U_1	$Q_{sf(3)} = Re_1U_3 + Re_2U_2 + Re_3U_1$
4	Re_1U_4	Re_2U_3	Re_3U_2	$Q_{sf(4)} = Re_1U_4 + Re_2U_3 + Re_3U_2$
5		Re_2U_4	Re_3U_3	$Q_{sf(5)} = Re_2U_4 + Re_3U_3$
6			Re_3U_4	$Q_{sf(6)} = Re_3U_4$

Notes: *In the table, for simplicity, Re and U are not written in italics.*

As seen from the illustration above, the linear hydrograph theory assumes that the total runoff response for any time period is a linear addition ('convolution') of the runoff responses from each *Re*.

Despite this the key problem of the UH method is to determine the UH shape and ordinates. These ordinates can be derived if we have measurements of Q_{sf} responses for a given rainfall event, but such data sets are rarely available in practice because discharge here is not the total discharge in the river at any given time but only that from the surface runoff.

Thus, in practice, a UH method is often derived using certain parameters that can be realistically estimated from catchment physical properties. One or more of the four such parameters are commonly used by different versions of unit hydrographs methods (see, e.g. Feldman, 2000; USDA, 2010b; and Singh et al., 2014). These parameters are the 'time of concentration' and 'time lag', and 'peak' and 'duration' of the UH. Note the difference between the time of concentration and time lag. The time lag refers to the time difference between the generation of rainfall excess in the catchment and runoff response at the outlet, and usually measured from the centroid of the *Re* verses time diagram to the peak of the hydrograph (Snyder, 1938). A slightly different method used in HBV model (Bergström, 1992) uses the duration of the UH as its parameter and works out the detailed transfer function ordinates by calibration assuming an equilateral triangular shape. In most of the available hydrological models the unit hydrograph parameters are treated as calibration parameters.

5.3.1.2 Linear reservoir

A more convenient method to implement in a model is the linear reservoir (LR) method, which is more generally a conceptual method with a broad range of applications in hydrological modelling. The LR method is based on two principles.

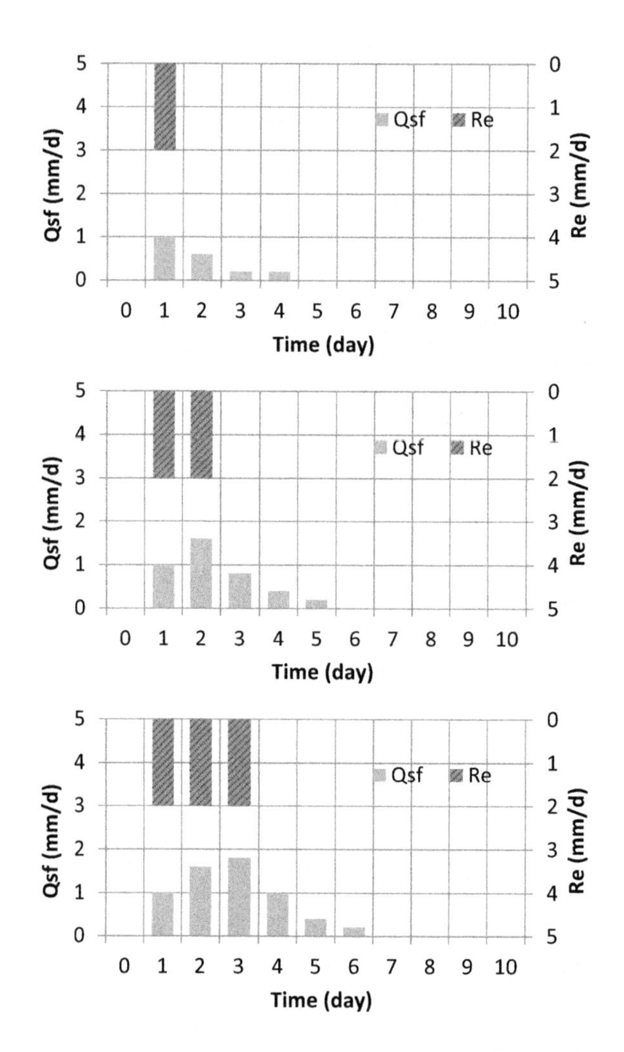

FIG. 5.2

Illustration of the UH method: Surface runoff (Q_{sf}) from 2 mm/day of rainfall excess (*Re*) in (1) one day (top), (2) two consecutive days (middle), and (3) three consecutive days (bottom). UH ordinates used are 0.5, 0.3, 0.1 and 0.1.

The first is the conservation of mass expressed in terms of the change in rainfall excess stored in the catchment within a certain time. That is

$$\text{Change in Storage } (S) \text{ over time } (t) = \text{Inflow } (Q_{\text{in}}) - \text{Outflow } (Q)$$

$$\frac{dS}{dt} = Q_{in} - Q \tag{5.1}$$

where S is the volume [L^3], and Q's are in [L^3T^{-1}].

The second is the assumption that tells that the storage (S) at any time is linearly related to the average outflow from the catchment (Q) within the same time period. Thus.

Storage (S) = Reservoir Constant (k) × Outflow (Q)

$$S = kQ \tag{5.2}$$

Using these two relationships and a simple numerical approximation of the differential equation, we can express the unit hydrograph ordinate or the outflow discharge $Q(t)$ at time t as a weighted sum of the inflow $Q_{in}(t)$ and the outflow of the previous time step, $Q(t-1)$. Where, the inflow is the rainfall excess for the same time period, that is $Q_{in}(t) = Re\ (t)$, and the outflow $Q(t)$ is the surface flow $Q_{sf}(t)$. Thus, we can write (see Chapter 6 for the derivation)

$$Q_t = wQ_{in,t} + (1-w)Q_{t-1} \tag{5.3}$$

where the weighting factor w, which controls the shape of the hydrograph depends on the relative values of the computation time step and reservoir storage constant k by

$$w = \frac{\Delta t}{k + 0.5\Delta t} \tag{5.4}$$

So k here is the only parameter for the LR method, which is typically determined by calibration. k has the same unit as time, but it is not the value of k alone but its value relative to the time step that is important. A larger k compared to Δt shifts the weight to $Q(t-1)$ and represents a slow transformation, while a smaller k shifts the weight towards Q_{in}. With $k = 1.5\Delta t$, the weights are equally divided to Q_{in} and $Q(t-1)$. When k is one half of Δt, the full weight is given to Q_{in}, meaning that all the Re available at a given time step reaches the stream in the same time period. Note that this is an exponential decay equation, so the discharge can continue forever (although with a small amount). So the algorithm must put a limit when to stop based on the volume conservation.

There is one issue with this numerical solution of the LR method, which is that it does not work when $k < \Delta t/2$. However, as we noted, when $k = \Delta t/2$, all the Re available over the time period Δt reach the stream in the same time period. So we can safely assume that this is also the case for $k < \Delta t/2$. This assumption allows us to rewrite the equation for weighting factor w, as

$$w = \begin{cases} \dfrac{\Delta t}{k + 0.5\Delta t} & \text{for } k \geq \Delta t/2 \\ 1 & \text{for} k < \Delta t/2 \end{cases} \tag{5.5}$$

In Fig. 5.3, three surface runoff (outflow hydrograph) graphs are plotted from the same intensity of rainfall for three consecutive days used in the UH example (Fig. 5.2) using different reservoir storage constants. In the top graph, the reservoir constant is one half of the time step. In the middle, the reservoir constant is equal to the time step, and in the bottom graph, it is one and a half times the time step. As you can see, with $k = 0.5\Delta t$, all the rainfall excess becomes runoff in the same time step. Whereas with $k = 1.5\Delta t$, the runoff response continues for more time periods.

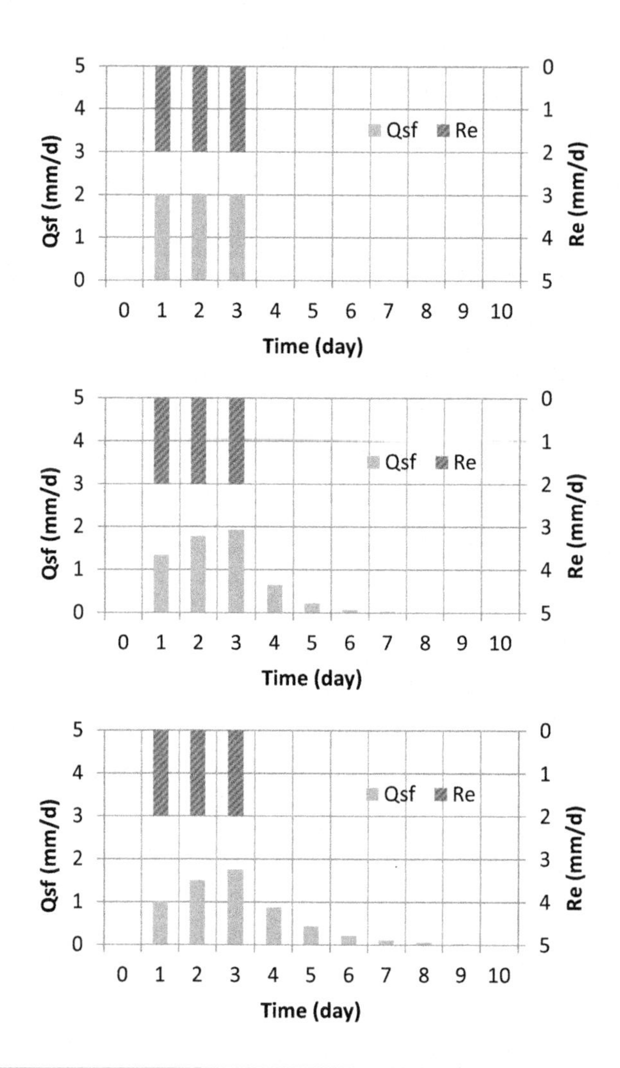

FIG. 5.3

Illustration of the linear reservoir method for surface flow routing with three reservoir constants: $k=0.5\Delta t$ (top), $k=\Delta t$ (middle) and $k=1.5\Delta t$ (bottom). The rainfall excess of 2 mm/d for three consecutive days are assumed.

5.3.2 Kinematic wave method

The kinematic wave equation is an approximation of the fully dynamic wave equation. Both the dynamic wave and kinematic wave equations are presented in Chapter 7 on river flow routing. When the 1D formulation of the kinematic wave equation is used it implies that the changes or dynamics of the flow is considered only in one primary direction of flow. In channel or river flow the primary direction is normally well defined, unless the channel is too wide, because it is aligned by its banks. In case of a surface or overland flow, the flow direction is often not well defined. Therefore, although the theory of the kinematic wave applied in surface flow routing and river flow routing is the same, it differs in the implementation as to how the catchment surface topography can be viewed as a channel.

The continuity equation for a 1D channel flow, which is also used with the kinematic wave method is given by Eq. (5.6) (see also Chapter 7). Note that the expression of the continuity equation for the channel flow differs from that defined in the linear reservoir method. That is because in the channel flow, the changes are expressed with respect to time as well as distance along the flow direction. Thus, the flow travel length is explicitly used here, whereas the length is unspecified in the LR method. As there is no length, there is no need for the channel cross-section and area, and that makes the implantation of the LR method a lot simpler than the kinematic wave method.

$$\frac{\partial Q}{\partial x} + \frac{\partial A}{\partial t} = q_t \tag{5.6}$$

where x is distance [L], A is the cross-section area of flow, and q_t is the lateral flow discharge per unit length in [$L^2\,T^{-1}$], which comes from the rainfall excess Re . The continuity equation is sometime expressed only with Q as

$$\frac{\partial Q}{\partial x} + \frac{1}{c}\frac{\partial Q}{\partial t} = q_t \tag{5.7}$$

where c is celerity [LT^{-1}] given by $c = dQ/dA$ (Chow et al., 1988). The above two Eqs. (5.6) and (5.7) are expressed in conservative and nonconservative forms, respectively. See Guinot (2008) for more on conservative and nonconservative forms of the equation. In the kinematic wave approximation, the momentum equation reduces to the condition that the friction force (proportional to the friction slope, S_f) = gravity force (proportional to the channel bed slope, S_0). This is the condition that the 'friction slope is parallel to channel bed slope' and leads to the derivation of the Manning's and Chezy's formulas.

Note that Eq. (5.6) can also be expressed using velocity V [LT^{-1}] and water depth h [L] instead of Q, by substituting $A = bh$ and $Q = AV = bhV$.

These equations are typically solved using a numerical method. One such method commonly used in free surface flow is called the finite-difference method. The finite-difference method can be implemented in various ways, e.g. forward difference, backward difference, etc. But they can be viewed as two key types: explicit and implicit. The explicit solutions are easy to implement and requires less computation time, but they work only within a limited range of $\Delta t/\Delta x$ ratio. When the Courant Number, which is defined as celerity (c) times $\Delta t/\Delta x$, is larger than 1, the explicit numerical scheme becomes unstable, meaning that any small numerical error amplifies at each time step making the computation senseless and brings to an end. The implicit scheme on the other hand involves iterations, one way or another, making it relatively complicated to implement and computationally time consuming.

As we can see in the continuity equation above, the unknown quantities Q and A are varying on space (x) and time (t). How they are varying in space and time

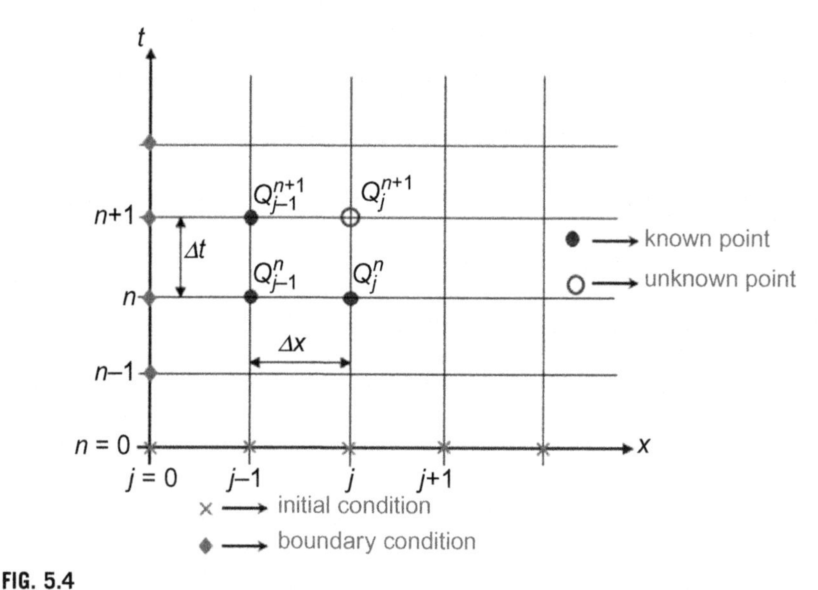

FIG. 5.4

Notations used for numerical space–time discretization.

can be best illustrated using a 'space–time diagram', such as shown in Fig. 5.4. In the horizontal axis, we put the measurement in space, which for 1D flow is the distance along the flow direction. In the vertical axis, we put time. Then we add grids of small intervals Δx and Δt in both axes. We use letter j to indicate a grid for space (length) and letter n to indicate a grid for time. If a particular grid is j, then for the next grid (by one step Δx) we write $j+1$ and the grid one step back we write j–1. Similarly, we write n, $n+1$ and n–1 for the time grids. On each grid points, we write the values of a particular quantity of interest, e.g. Q or h. Which means if we can fill up all the grids with values of Q, for example, then we know how Q is changing over space and time throughout the model domain. We use letters j and n as subscript and superscript to refer to a variable (e.g. Q) in a particular grid. For example, to represent a discharge at space grid j and time grid n, we write Q_j^n, and so forth. Note that for simplicity, when used inside text, j and n are written in parentheses instead of subscript and superscript. Thus, Q_j^n is written inside the text as Q (j, n).

The question is of course how to fill up all the grids. In order words, how can we calculate values of, say, Q for throughout the length and time? The main idea here is to estimate the value for an unknown grid point from at least two or more known neighbouring grid points. Which means we need to know the values for some grid points to be able to start the computation. These values come from what we call initial and boundary conditions. An initial condition is the values of the dependent variable at the beginning of the computation (i.e. at the time grid n for $t=0$) at every section in space (i.e. for all j's). This is basically all the grids at the lower most row in the space–time plot (Fig. 5.4). Similarly, a boundary condition is the value of the dependent variable for all time steps of the computation (i.e. for all n's) at the beginning of the space grid (i.e. grid j for $x=0$). This is called the upstream boundary condition, and depending on the type of the equation to be solved, we may also need a downstream boundary, which is for the final space grid at all times of the computation. Just as the initial condition, the upstream boundary condition fills up all the grid points at the left most column in the space–time diagram. Thus, the initial and boundary conditions provide necessary starting values to compute the rest of the grids using a numerical scheme.

Now let us look at how different schemes can be formed. In Fig. 5.4, an unknown point is marked with a circle, neighbouring known points are shown with a filled circle. Note also the notations of j and n with the Q in the figure. The unknown grid point that is to be estimated next is indicated by j and $n+1$. Meaning that if we are at grid j now, we want to estimate how much will be the discharge at this section (indicated by grid j) after one-time step from now, that is at $n+1$. Note that you may find some variations in the literature in the way these notations are used for the known and unknown points for space (distance) and time.

Based on our notation, the value for grid $(j, n+1)$ can be computed using two known points, e.g. $(j-1, n)$ and (j, n) or $(j-1, n)$ and $(j-1, n+1)$; three points, e.g. $(j-1, n)$, (j, n) and $(j-1, n+1)$ or $(j-1, n)$, (j, n) and $(j+1, n)$. Thus, depending on which points are used the numerical schemes differ and can be explicit or implicit.

Three examples are illustrated here: two explicit methods with two known points and one implicit with three known points.

(1) Explicit with three points with Q

$$\frac{Q_j^n - Q_{j-1}^n}{\Delta x} + \frac{1}{c}\frac{Q_j^{n+1} - Q_j^n}{\Delta t} = q_t \tag{5.8}$$

The kinematic wave celerity, c, can be approximated using the following relationship between the cross-sectional area (A) of the channel and the discharge (Q):

$$A = \alpha Q^{\beta} \tag{5.9}$$

The parameters α and β are related to the channel shape and type of the steady-uniform formula (e.g. Manning or Chezy equations) used (see, e.g. Chow et al., 1988; Feldman, 2000). For example, for Manning's formula, β equals to 0.6 and α relates to the roughness (n), wetted perimeter (P) and channel bed slope (S_o) given by

$$\alpha = \left(\frac{nP^{0.667}}{S_o^{0.5}}\right)^{0.6} \tag{5.10}$$

(2) Explicit with three points with Q and A

$$Q_j^{n+1} = Q_{j-1}^{n+1} + q_t \Delta x - \frac{\Delta x}{\Delta t}\left(A_{j-1}^{n+1} - A_{j-1}^{n}\right) \tag{5.11}$$

(3) Implicit with four points

$$\frac{\theta\left(Q_j^{n+1} - Q_{j-1}^{n+1}\right) + (1-\theta)\left(Q_j^n - Q_{j-1}^n\right)}{\Delta x} + \frac{\left(A_j^{n+1} - A_j^n\right) + \left(A_{j-1}^{n+1} - A_{j-1}^n\right)}{2\Delta t} = q_t \tag{5.12}$$

where θ is the weighting factor with a value between 0.5 and 1.0, and may be considered as a calibration parameter. This is the same numerical scheme used by Alley and Smith (1982). Rearranging all the unknown terms on the right-hand side, we obtain

$$\theta Q_j^{n+1} + \frac{\Delta x}{2\Delta t} A_j^{n+1} = \Delta x q_t + \theta Q_{j-1}^{n+1} - (1-\theta)\left(Q_j^n - Q_{j-1}^n\right) + \frac{\Delta x}{2\Delta t}\left(A_j^n + A_{j-1}^n - A_{j-1}^{n+1}\right) \quad (5.13)$$

In the four points (implicit) method, there are two unknowns on the left-hand side, Q and A, at $(j, n+1)$. All the terms on the right-hand side are either for one time level earlier (i.e. n) or one section upstream (i.e. $j-1$), so these are known by the time we estimate for $(j, n+1)$. Because there are two unknowns on the left-hand side, we still need to use the A-Q relationship (Eq. 5.9) here to replace Q by A first. By doing that the finite difference equation can be rearranged into a quadratic equation for the unknown A $(j, n+1)$ as illustrated by Alley and Smith (1982). This type of equation is also commonly solved using some iterative procedures or numerical methods.

5.3.3 Diffusion wave method

A more physically based model often used in surface flow routing is the diffusive wave model in the 2D formulation. Likewise the kinematic wave model, the momentum equation for the diffusive wave model is also derived from the Saint Venant equation with some approximation. While the acceleration terms (convective and local) and the pressure term are neglected in the kinematic wave approximation, only the acceleration terms are neglected in the diffusive wave approximation. Thus, the diffusive wave model in 1D is represented by Eq. (5.14) (also shown in Chapter 7).

$$S_f = S_0 - \frac{\partial h}{\partial x} \quad (5.14)$$

To further simplify this, take z as the bed or ground surface level (i.e. the ground surface elevation represented by the grid cell) and H as the water surface level, i.e. $H = z + h$, then we obtain

$$S_f = -\frac{\partial z}{\partial x} - \frac{\partial (H - z)}{\partial x} \quad (5.15)$$

Note that the bed slope term is negative ($-\partial z/\partial x$), because downward bed slope is considered positive. Finally, the momentum equation for diffusive wave approximation in 1D is

$$S_f = -\frac{\partial H}{\partial x} \quad (5.16)$$

To write the continuity equation in 2D application, we need to add the change in variable of interest (e.g. Q or h) with respect to change in distance in y-direction as well. Thus the continuity equation becomes

$$\frac{\partial h}{\partial t} + \frac{\partial q_x}{\partial x} + \frac{\partial q_y}{\partial y} = R_e \quad (5.17)$$

where q_x and q_y are the discharge per unit width [L^2T^{-1}] in the x-axis direction and y-axis direction, respectively, and Re is the rainfall excess rate averaged over the given time step.

For the momentum Eq. (5.16) for 2D we can write,

$$S_{f\_x} = -\frac{\partial H}{\partial x} \tag{5.18}$$

$$S_{f\_y} = -\frac{\partial H}{\partial y} \tag{5.19}$$

In the form of Manning's equation, the S_f is approximated by

$$S_f = \frac{n^2 V^2}{h^{\frac{4}{3}}} \tag{5.20}$$

Then using q_x and q_y with $q = hV$, we can write

$$S_{f\_x} = \frac{n^2 q_x^2}{h^{\frac{10}{3}}} \text{ and } S_{f\_y} = \frac{n^2 q_y^2}{h^{\frac{10}{3}}} \tag{5.21}$$

Finally, replacing S_{f_x} and S_{f_y} in Eqs. (5.18) and (5.19), we obtain

$$q_x = \frac{(h_{av})^{\frac{5}{3}}}{n} \left(-\frac{\partial H}{\partial x} \right)^{0.5} \tag{5.22}$$

$$q_y = \frac{(h_{av})^{\frac{5}{3}}}{n} \left(-\frac{\partial H}{\partial y} \right)^{0.5} \tag{5.23}$$

Note that the water depth used to calculate q is the average depth that is maintained during the time step, and so it is represented by h_{av} in the above equation. Moreover, it is the same average depth (h_{av}) for both q_x and q_y. So basically, we need to solve the three Eqs. (5.17), (5.22) and (5.23) for each grid cell. These equations may look relatively simple, but solutions of them are challenging, even with an explicit scheme. Therefore, they are usually solved using some kind of approximations (see e.g. DHI, 2017a). A more detailed procedure of the 2D diffusive wave method for surface flow routing can be found in DHI (2017a).

5.4 How do different catchment models treat surface flow routing?

Brief descriptions of methods and approaches used in 17 different hydrological models for surface flow routing are presented here (Table 5.2). The selected models are among the widely used models, but the list is not exhaustive, and the purpose of presenting the table is not for giving the author's judgement about the models. It is simply intended to describe the key concepts of the methods and approaches used in

Table 5.2 Surface flow routing and transfer methods used in various hydrological models.

Modelling software	Method available	Additional information
BROOK90	It does not have surface flow routing. It's a lumped model without horizontal transfer of water, so no surface and river flow as such (Federer, 2002).	
CASC2D	Surface flow routing is based on 2D diffusive and fully dynamic equations. Flow from a grid-cell can be routed to four possible cells (in four directions) depending on water surface elevation differences (Ogden, 2001).	
CHARM (also WATFLOOD)	It uses the Manning's uniform flow formula analogy, but the roughness parameter represents more than just the flow resistance in Manning's formula, whose value largely relies on calibration (Kouwen, 2018, for surface routing see p. 2–14).	CHARM: Canadian Hydrological and Routing Model
Flo2D	Surface (overland) flow routing is based on the 2D diffusive wave equation. Flow from a grid-cell can be routed to eight neighbouring cells proportional to velocities calculated individually to each possible cells (FLO-2D, 2003).	
HBV (and HBV Light)	It does not have the (direct) surface flow routing as such. Runoff components are from the soil layer and groundwater layer(s), which are combined together before applying a transfer factors. The factors are defined using a symmetrical (equilateral) triangle with its base (duration) as a calibration parameter (Bergström, 1992; Seibert, 2005).	
HEC-HMS	It has unit hydrograph (different versions) method and kinematic wave method in 1D formulation (Feldman, 2000).	

Table 5.2 Surface flow routing and transfer methods used in various hydrological models. —*cont'd*

Modelling software	Method available	Additional information
LISFLOOD	It uses the kinematic wave method to route grid cell by cell along a defined flow direction based on steepest descent (Burek et al., 2013, see p. 24 for the routing method and p. 127 for flow direction).	
MIKE SHE	It has 2D diffusive wave method and a simplified conceptual method used in the Stanford Watershed Model (Crawford and Linsley, 1966), which uses the continuity equation and Manning's uniform flow analogy to estimate ponding storage in a surface area and discharge from the area (DHI, 2017a).	
NAM	It uses the linear reservoir transfer function with a time coefficient (DHI, 2017b).	
PCR-GLOBWB	It does not have surface flow routing as such, but adds the surface flow volume to the river cell without time delay (Van Beek and Bierkens, 2009).	This description is based on PCR-GLOBWB 1.0, and currently version 2.0 is also available https://globalhydrology.nl/research/models/pcr-globwb-2-0/.
PRMS (ver. 4) also used in GSFLOW	It routes the surface flow from an upslope hydrological response unit (HRU) to any number of downslope HRUs following a path called 'directed, acyclic-flow network' with different percentages of the contributing area (Markstrom et al., 2015; Markstrom et al., 2008).	PRMS: Precipitation Runoff Modelling System (Markstrom et al., 2015). GSFLOW: Coupled Ground-water and Surface-water Flow Model (Markstrom et al., 2008)
RRI model	It uses the 2-D diffusion routing (Sayama, 2017).	
SWAT	It takes a fraction of the surface runoff generated in the catchment as the surface flow contribution to the river flow the fraction varies exponentially depending on the ratio of a parameter 'surface lag' and 'time of concentration' (Neitsch et al., 2011).	

Continued

Table 5.2 Surface flow routing and transfer methods used in various hydrological models. —*cont'd*

Modelling software	Method available	Additional information
UBC model	It uses the unit hydrograph with two parameters (Quick and Pipes, 1977).	UBC: University of British Colombia
VIC	It uses the unit hydrograph method with a triangular shape (Liang et al., 1994; Gao et al., 2010).	VIC: Variable Infiltration Capacity. See also the VIC version 5 webpage: https://vic.readthedocs.io/en/master/
WaSiM	It uses the kinematic wave formula with the Manning's equation, and the flow is routed cell by cell along the steepest gradient direction (with or without water depth consideration in slope estimation) with a possibility of multiple flow direction (Schulla, 2012).	
Xinanjiang model	No routing method is used for the surface flow in the original version. The surface flow is added directly to the channel flow without modification (Zhao, 1992).	

the models. The descriptions attempt to capture the main essence or features, but may not be detailed enough to provide complete details of the methods, for which the referenced literature should be consulted.

References

Alley, W.M., Smith, P.E., 1982. Distributed Routing Rainfall-Runoff Model, Version II. Open-File Report 82–344. US Geological Survey, Reston, Virginia 22,092. https://doi.org/10.3133/ofr82344.

Bergström, S., 1992. The HBV model—its structure and applications. SMHI RH No 4. Norrköping.

Burek, P., Van der Knijff, J., De Roo, A., 2013. LISFLOOD distributed water balance and flood simulation model, revised user manual. JRC Technical Report 'EUR 26162 EN'.

Chow, V.T., Maidment, D.R., Mays, L.W., 1988. Applied Hydrology, International Edition. McGraw-Hill, Singapore.

DHI (2017a). MIKE SHE Volume 2: Reference Guide. DHI, Denmark. https://manuals.mikepoweredbydhi.help/2017/Water_Resources/MIKE_SHE_Printed_V2.pdf; Accessed on 31 July 2021.

DHI (2017b). MIKE SHE Volume 1: User Guide. DHI, Denmark. https://manuals.mikepoweredbydhi.help/2017/Water_Resources/MIKE_SHE_Printed_V1.pdf; Accessed on 31 July 2021.

Crawford, N.H., Linsley, R.K., 1966. Digital Simulation in Hydrology: Stanford Watershed Model IV. Technical Report No. 39. Department of Civil Engineering, Stanford University, Stanford, California.

Federer, C.A., 2002. BROOK 90: A simulation model for evaporation, soil water, and streamflow. http://www.ecoshift.net/brook/brook90.htm.

Feldman, A.D. (Ed.), 2000. Hydrologic Modelling System HEC-HMS Technical Reference Manual. US Army Corps of Engineers, Hydrologic Engineering Centre, Washington, DC.

FLO-2D, 2003. FLO-2D User Manual. Nutrioso.

Gao, H., Tang, Q., Shi, X., Zhu, C., Bohn, T.J., Su, F., Sheffield, J., Pan, M., Lettenmaier, D.P., Wood, E.F., 2010. Water Budget Record from Variable Infiltration Capacity (VIC) Model. In: Algorithm Theoretical Basis Document for Terrestrial Water Cycle Data Records. UNSPECIFIED. Available for download from https://eprints.lancs.ac.uk/id/eprint/89407.

Guinot, V., 2008. Wave Propagation in Fluids: Models and Numerical Techniques. ISTE Ltd., London.

Kouwen, N., 2018. Canadian Hydrological And Routing Model. User manual. ENVIRONMENT CANADA. Also see: http://www.civil.uwaterloo.ca/watflood/index.htm.

Liang, X., Lettenmaier, D.P., Wood, E.F., Burges, S.J., 1994. A simple hydrologically based model of land surface water and energy fluxes for general circulation models. J. Geophys. Res. 99 (D7), 14415–14428.

Markstrom, S.L., Niswonger, R.G., Regan, R.S., Prudic, D.E., Barlow, P.M., 2008. GSFLOW–Coupled Ground-water and Surface-water FLOW model based on the integration of the Precipitation-Runoff Modeling System (PRMS) and the Modular Ground-Water Flow Model (MODFLOW-2005): U.S. Geological Survey Techniques and Methods 6-D1. 240 p.

Markstrom, S.L., Regan, R.S., Hay, L.E., Viger, R.J., Webb, R.M.T., Payn, R.A., LaFontaine, J.H., 2015. PRMS-IV, the Precipitation-Runoff Modeling System, Version 4, Techniques and Methods 6–B7. U.S. Geological Survey, Reston, Virginia.

Neitsch, S.L., Arnold, J.G., Kiniry, J.R., Williams, J.R., 2011. Soil and Water Assessment Tool Theoretical Documentation Version 2009. Texas Water Resources Institute. Available electronically from https://hdl.handle.net/1969.1/128050.

Ogden, F.L. (1998, rev 2001). A Brief Description of the Hydrologic Model CASC2D. Univ. Connecticut.

Quick, M.C., Pipes, A., 1977. U.B.C. Watershed Model. Hydrol. Sci. J. 22 (1), 153–161.

Sayama, T., 2017. Rainfall-Runoff Inundation (RRI) Model, Version 1.4.2. International Centre for Water Hazard and Risk Management (ICHARM) and Public Works Research Institute (PWRI), Japan.

Schulla, J., 2012. Model Description WaSiM. Hydrology Software Consulting J. Schulla, Zurich. pp. 300.

Seibert, J., 2005. HBV Light Version 2 User's Manual. Department of Physical Geography and Quaternary Geology, Stockholm University.

Singh, P.K., Mishra, S.K., Jain, M.K., 2014. A review of the synthetic unit hydrograph: from the empirical UH to advanced geomorphological methods. Hydrol. Sci. J. 59 (2), 239–261. https://doi.org/10.1080/02626667.2013.870664.

Snyder, F.F., 1938. Synthetic unit-graphs. Trans. Am. Geophys. Union 19 (1), 447–454. https://doi.org/10.1029/TR019i001p00447.

USDA, 2010a. National Engineering Handbook, Part 630 Hydrology, Chapter 15. United States Department of Agriculture, Natural Resources Conservation Services.

USDA, 2010b. National Engineering Handbook, Part 630 Hydrology, Chapter 16. United States Department of Agriculture, Natural Resources Conservation Service.

Van Beek, L.P.H., Bierkens, M.F.P., 2009. The Global Hydrological Model PCR-GLOBWB: Conceptualization, Parameterization and Verification, Report. Department of Physical Geography, Utrecht University, Utrecht, The Netherlands. http://vanbeek.geo.uu.nl/suppinfo/vanbeekbierkens2009.pdf.

Zhao, R.-J., 1992. The Xinanjiang model applied in China. J. Hydrol. 135, 371–381.

CHAPTER

Models of groundwater (saturated zone) flow

6

6.1 Role of groundwater flow in a catchment model

As discussed in the previous chapter, water in a catchment reaches the river system broadly in two ways: as surface flow and subsurface flow. In Chapter 5, we dealt with the surface flow: its conceptualization and routing methods. Once the precipitated water infiltrates into the soil it forms a part of the subsurface flow, which is divided into unsaturated and saturated (or groundwater) zones (see Chapter 4). Subsurface flow is the water that travels underneath the land surface in the porous soil, and where the river water levels are below the saturated subsurface level or water table, the subsurface flow feeds to the rivers. Water in the unsaturated zone may also reach the streams through interflow, but it is primarily the groundwater flow (also called baseflow) that subsurface flow contributes to the river flow. In Chapter 4 we dealt with how water moves in the unsaturated layer of the subsurface and the interflow to the river. The soil layer underneath the unsaturated zone with all of the effective porosity filled with water is the saturated zone. Such a saturated layer that also permits water to travel through it is called an aquifer, which is often a major source of water for many river systems, particularly during the dry period of the year. The aquifer that is directly underneath the unsaturated soil layer is called an unconfined aquifer, because it is a kind of free aquifer such that the pressure at the top of the aquifer, which is also called the groundwater table, is at atmospheric pressure. Another name for the unconfined groundwater table is phreatic surface. Note that a relatively thin layer, called a capillary fringe, exists between the water table of the unconfined aquifer and the unsaturated layer above it. This layer is also in saturation due to capillaries that with the suction force pull water from the saturated aquifer to the pores of the soil above it. The height of this layer depends on the suction force acting at the location. In general, smaller the soil grains larger the force and consequently larger the capillary fringe. The capillary fringe is considered part of the vadose zone.

The confined aquifer is formed in between impervious layers. The impervious layers can be an aquitard or aquiclude. An aquitard is a formation with very low permeability, which can transmit water vertically, e.g. from one aquifer layer to another, but hardly transmits water horizontally. An aquiclude is a formation with practically zero permeability, and transmits water neither horizontally nor vertically (Bouwer, 1978). Which means that the recharge to the confined aquifer takes place from areas at some higher elevation in the catchment where it may be in contact with unsaturated soil layer, or at places where the aquifer above is leaking water through an

Catchment Hydrological Modelling. https://doi.org/10.1016/B978-0-12-818337-3.00002-7

aquitard. Unlike in the unconfined aquifer, pressure higher than atmospheric prevails in a confined aquifer. Thus the water flow in a confined aquifer is analogous to a pipe flow but of course with a lot more resistance. The unconfined and confined aquifers are also commonly referred to as shallow and deep aquifers, respectively.

The unconfined or shallow aquifer usually supplies water to streams in the catchment. Flow from a deep aquifer often discharges to a river system further downstream from the catchment. But some confined aquifers, depending on the depth to the aquifer and topography of the catchment, may also drain to streams in the same catchment as well as streams in downstream catchments. This aspect of which river system an aquifer discharges and from where it is recharged is important for understanding water balance of a catchment. If an aquifer is recharged from precipitation falling on one catchment, say catchment U, and it supplies water to a river system in another catchment downstream, say catchment D, then for catchment U, the amount of recharge to the aquifer is a loss. Whereas for catchment D, the groundwater flow from that aquifer is input from an external source. We will refer to this again when we discuss groundwater water balance in Section 6.3. Aquifers may also recharge from and discharge to lakes or reservoirs.

The part of a river flow that is discharged from aquifers (both confined and unconfined) is called a baseflow, Q_{bf}, or more directly groundwater flow, Q_{gw}. Note that groundwater flow in general may also be called the subsurface flow, which includes the flow from saturated zone (aquifer) and interflow, Q_{if}, from the unsaturated zone. The mechanism of the interflow is discussed in Chapter 4. As discussed in Chapter 5 the surface flow (or direct runoff) takes place only on some instances, that is when the rainfall intensity exceeds the infiltration capacity or rainfall occurs when the top soil is already saturated. So direct runoff is not an every day's event and it supports the river flow only limited and intermittent periods in a year. However, many river systems are perennial, so clearly interflow and baseflow play substantial role to supply water to the rivers most of the time. In the river system in glacierized catchments or catchments with snow precipitation, snow/glacier melt water is another source that keeps rivers flowing during dry periods. Even so, some or large part of snow melt water is more likely to be infiltrated before reaching the rivers. So again the melt water becomes part of interflow and baseflow. Also, the interflow usually does not last very long after a rainfall event, so for the dry season or most part of it, it is either the groundwater baseflow or snow/glacier melt that keeps the rivers flowing. It is therefore inevitable to have a groundwater flow model component within a catchment hydrological model.

In Fig. 6.1, the connections of the groundwater flow model to other components of the catchment model are shown. The connection diagram for the groundwater flow model is rather simple. Basically it is connected to the unsaturated zone for recharge (connection 1) and capillary flow from the unconfined aquifer to the unsaturated (vadose) zone (connection 2), and to the river flow model for baseflow supply to the river (connection 3) and recharge from the river to the aquifer (connection 4). Note that lake or reservoir models are not presented here. Aquifers (groundwater) may also be connected to lakes and reservoirs for recharge/discharge.

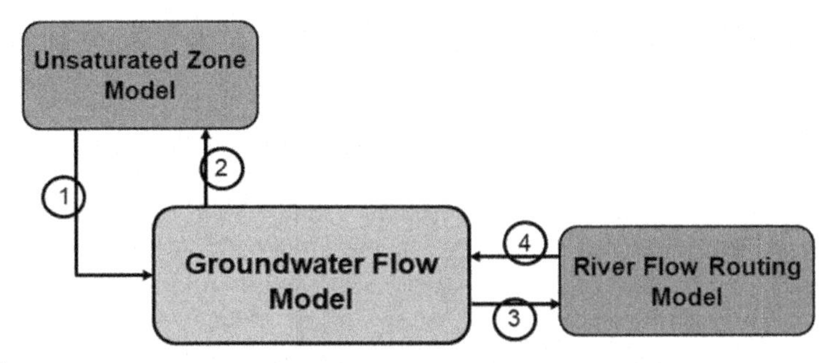

FIG. 6.1

Connections of the Groundwater Flow Model to other components of the catchment model. Connections indicated by the numbers represent: (1) Recharge input to the groundwater, (2) Moisture transport through capillary rise from the SZ (unconfined aquifer) to UZ, (3) Q_{gw} to the river flow (when $h_{riv} < h_{aq}$), and (4) Recharge to groundwater through river beds (when $h_{riv} > h_{aq}$). Connections between components other than directly with the groundwater flow model are not shown here.

6.2 Conceptualization of groundwater system in a catchment

Groundwater exists in a complicated flow system and our understanding of it is hindered partly because we do not see it as we do river flow. How deep are the aquifers? From where to where they extend? Are they contained within the same catchment boundary or extend beyond? How thick are they and how much water they may be holding and how good are they to permit water to flow through? Where are they recharged from and at what rate? Which streams/rivers are they connected to for discharge/recharge? These are the primary questions we need to know to accurately model groundwater flow contribution to a river system. There exists models only for groundwater, which are sophisticated and physically based. I call them a 'specialist' model, in this case, for groundwater modelling. The popular MODFLOW (Harbaugh et al., 2000) is one example. For practical applications, groundwater models with the modelling software like MODFLOW can be built with the data indicated above about the groundwater system. However, groundwater does not exist in isolation. It is intrinsically connected to the layer above it for recharge and to the rivers for flow exchange, as we have seen in Fig. 6.1. So the groundwater flow must be included as a component in a hydrological catchment model. In practice when catchment hydrological models are built, a lot of the data about groundwater in the catchment are usually unavailable. That is often the case when a catchment model covers a large area. Moreover many of the hydrological models do not have the groundwater flow model components that are capable of utilizing the detailed information about aquifers to simulate the flow in the groundwater system. To model a groundwater system with the detailed representation of its size, location, aquifer properties, etc. we would need a physically based 2D/3D model as is the case in the 'specialist' groundwater models. We discuss about equations for such a model formulation in Section 6.4.2. With the exception of few hydrological models, such as Mike SHE, such a possibility is unavailable in hydrological models.

In dealing with groundwater flow within a catchment hydrological model, hydrologists have managed to tackle the complicated groundwater system with rather simple, efficient and often effective methods. Contrary to the 2D/3D groundwater model, the groundwater flow component within a hydrological model is usually lumped-conceptual (0-dimensional, meaning only the time series of inflow, outflow and volume change is represented). This simplicity used in a catchment model to simulate groundwater is just as 'specialist' groundwater models may do to simulate surface water component necessary for the groundwater. Few hydrological models try to take it one step further to translate the groundwater volume to groundwater table (height), but the computed height is not intended to represent actual groundwater table or groundwater head. See, e.g., the groundwater component in the Soil and Water Assessment Tool, SWAT (Neitsch et al., 2009), a popularly used hydrological model.

A commonly used conceptualization of groundwater system adopted in catchment models is to assume a shallow (unconfined) aquifer and a deep (confined) aquifer in each subcatchment without specification of their actual shape and location. It is like an imaginary object (aquifer), which stores water, receives recharge from the UZ, and discharges to a river at a certain rate. Then, to estimate groundwater discharge or baseflow (Q_{gw}), it assumes a functional relationship between the recharge (inflow), discharge (outflow) and change in the stored volume. We will discuss the principle and mathematics of such a relationship in the next section. In principle, the number of aquifers (or groundwater layers) does not have to be only two. Some hydrological models add a third layer—a further deeper aquifer. On the other hand, some models make it simpler and work with only one UZ and one SZ layer, and call them a fast reservoir and slow reservoir. Either way, the key intention in dealing with groundwater flow model within a catchment model is to try and simulate a realistic baseflow with only minimum data about the groundwater system.

6.3 Groundwater water balance and inflows-outflows

In earlier chapters, we discussed about water balance (conservation of mass) in a hydrological model, and we said that water balance equation can be written for any component of the catchment model, e.g. the soil layer, groundwater, river, etc. Here we write the water balance equation for the groundwater component. To recall, the basic form of the water balance equation is Change in Storage (volume per time) = Inflow − Outflow. The only thing that differs, to write water balance for different components, is the sources or forms of the inflow and outflow. For groundwater (aquifer) in a catchment, we can write:

Change in storage over time (dS/dt) =
+ Recharge to the groundwater aquifer (W_{rcg})
− Groundwater flow to UZ through capillary rise (W_{cap})

– Groundwater leakage to another aquifer (W_{leak})
– Groundwater outflow to river (Q_{gw})
+ Groundwater inflow from an aquifer outside the catchment (Q_{in_ext})
– Groundwater outflow to an aquifer/river/lake outside the catchment (Q_{out_ext}).

So, we write

$$\frac{dS}{dt} = W_{rcg} - W_{cap} - W_{leak} - Q_{gw} + Q_{in\_ext} - Q_{out\_ext} \tag{6.1}$$

where S is volume [L^3], t is time [T] and all other quantities are discharge rates in volume per time [$L^3\ T^{-1}$]. However, in practice, quantities such as recharge may be conveniently expressed as depth of water per time [LT^{-1}] (usually in mm/hour or mm/day). In that case, it must be multiplied by the recharge area [L^2], to keep all the terms in volume per time. Alternatively, all the volumes may be represented in depth [L], and when necessary multiply by their respective areas to get the volume in [L^3] and discharge rates in [$L^3\ T^{-1}$].

In the above equation, W_{rcg} is the input from the catchment precipitation to the groundwater. In case of an unconfined aquifer, recharge is assumed through the UZ, which is directly above in contact with the aquifer. Usually the flux from the aquifer to the soil layer through the capillary (W_{cap}) is small compared to the recharge and for simplicity may be neglected or combined with the recharge term and call it the net recharge. In case of a confined aquifer, the recharge may come from the high upstream area or from leaky parts of the layer separating the unconfined and confined aquifers (Bear, 1972; Bouwer, 1978). The leakage (W_{leak}) (that is, water transmitting from one aquifer to another) can be from confined to unconfined or vise versa depending on the head difference between the two. For a 'gaining' aquifer, W_{leak} becomes a part of the recharge (W_{rcg}). The last two terms (Q_{in_ext} and Q_{out_ext}) are the lateral 'in' and 'out' flows to the aquifer from and to the sources outside the catchment boundary. If the aquifer water balance is entirely contained within the catchment boundary, the Q_{in_ext} and Q_{out_ext} will be zero. Thus, assuming the W_{cap} and W_{leak} (if exists) are included in W_{reg} and the aquifer is contained entirely in the catchment (i.e. no Q_{in_ext} and Q_{out_ext}), the groundwater water balance equation becomes

$$\frac{dS}{dt} = W_{rcg} - Q_{gw} \tag{6.2}$$

In a conceptual catchment model with two groundwater layers, the recharge to the lower layer is commonly assumed through the upper layer. The groundwater discharge to the river in the catchment (Q_{gw}) is the primary interest in a catchment model.

In this water balance equation, the groundwater storage is represented as a single volume without specifying storage variation within the aquifer. A catchment may have more than one distinguishable aquifer, and water storage in the aquifer may vary spatially within the same aquifer. So, in that sense, the above water balance equation is a lumped representation of groundwater. Many hydrological models use a lumped representation of groundwater in a catchment as discussed earlier, but there are also models that use a grid-based distributed representation of catchment hydrological components including groundwater. In that case the water balance equation is written for each grid cell (or grid block) instead of the entire aquifer. Because 'in' and 'out' from a cell can vary so do the change in storage in the cell at each time step, and as an aquifer is a three-dimensional object, we can convert the change in volume to change in water level (h) so that we can keep track of water level variations within the aquifer cell by cell. Thus, the water balance equation for a grid cell of an aquifer can be represented as (McDonald and Harbaugh, 1984, p. 12)

$$\sum Q_i = S_s \frac{dh}{dt} \Delta V \tag{6.3}$$

where ΔV is the cell volume [L^3], the left-hand side is the sum of flow rates [$L^3 T^{-1}$] of all inflows (plus) to and outflows (minus) from a cell. On the right-hand side, the term S_s is *specific storage* (also called *specific storativity*, e.g. in Bear, 1972, p. 214). The *specific storage* is defined as the volume of water released from or added to the aquifer (or aquifer grid cell) per unit volume of the aquifer (or aquifer grid cell) per unit increase or decrease in hydraulic head in the aquifer (or aquifer grid cell). Thus, the unit of S_s is $m^3/m^3/m$ or m^{-1} [i.e. L^{-1}].

Note that the specific storage is different from *storage coefficient* (S_{co}) (also called *storativity*, e.g. in Bear, 1972 and Nonner, 2015) in that, the latter is defined as the volume of water that can be released from or added to per unit horizontal area of an aquifer per unit increase or decrease in hydraulic head. Thus, the storage coefficient is dimensionless (because $m^3/m^2/m$) and related to the specific storage and aquifer thickness (H) by

$$S_{co} = HS_s \tag{6.4}$$

where H is the aquifer thickness [L]. For an unconfined aquifer the term *specific yield* (S_y) is used in place of the storage coefficient. The distinction is because of the mechanism with which water is stored or released in confined and unconfined aquifers. In a confined aquifer, the change in the water volume (added or released) is attained primarily by the elasticity of the aquifer porous matrix, essentially by changing aquifer porosity due to the change in pressure (Bear, 1972; Nonner, 2015). While in the unconfined aquifer, when water is added or released the phreatic surface level changes by filling in or draining out water from the porous space.

6.4 Groundwater flow modelling methods

To understand a groundwater flow model and how different hydrological models handle groundwater, it is convenient to divide it into three components. First is the water balance of groundwater, because it tells us how the groundwater is playing a role in the catchment hydrology by showing the sources of inflows to and outflows from it. Second is the inflow and outflow processes and how they can be represented (mathematically) in the model. Third is about how the water is distributed within the groundwater and how it redistributes as inflow/outflow takes place continuously. We already discussed about the water balance and the major inflow/outflow components in Section 6.3. In the following sections, we present the groundwater models as two major types: conceptual and physically based. While describing the conceptual and physically based models we will discuss how these models represent the inflow/outflow components and (re)distribute groundwater in the aquifer, or keep track of the groundwater storage change in case of lumped conceptual models.

6.4.1 Linear reservoir method for groundwater flow modelling

The linear reservoir and exponential recession are two commonly used techniques to represent groundwater flow in catchment hydrological models. Although often they are referred to by different names ('liner reservoir' and 'exponential recession'), actually both are derived from the same simple assumption, which is that the reservoir storage volume, S, at any time t is linearly proportional to the outflow rate, Q_{out}. So, we can write: Groundwater storage (S) = Reservoir constant (k) × Groundwater outflow (Q_{out}). That is

$$S = kQ_{out} \tag{6.5}$$

We can see that the unit of the liner reservoir constant k has the unit of time (usually expressed in hour or day), because Q_{out} is in [$L^3 T^{-1}$] and S in [L^3]. Note that in the water balance Eq. (6.1 and 6.2) we used Q_{gw} for Q_{out} to specifically refer to the output to a river or stream.

Actually, the starting point for the liner reservoir is the continuity, which is the same for the water balance equation. We also applied the liner reservoir technique in Chapter 5 for surface flow routing. We can write the continuity as

$$\frac{dS}{dt} = Q_{in} - Q_{out} \tag{6.6}$$

This is the same as the simplest groundwater water balance Eq. (6.2) with inflow = recharge (i.e. $Q_{in} = W_{rcg}$) and outflow = baseflow (i.e. $Q_{out} = Q_{gw}$).

Using the symbols as in the groundwater water balance equation and substituting S with kQ_{gw} (which is kQ_{out} from Eq. 6.5), we obtain

$$\frac{d\left(kQ_{gw}\right)}{dt} = W_{rcg} - Q_{gw} \tag{6.7}$$

To derive a solution for Eq. (6.7), we consider two cases: one with the recharge term (W_{rcg}) and one without.

6.4.1.1 Case with no recharge

Let us consider a condition when there is no additional recharge to the groundwater during simulation. In a catchment such a condition (that is $W_{rcg} = 0$) exists when there is no rainfall for some time and the water content in the soil layer is already low to provide any more recharge. The linear reservoir Eq. (6.7) becomes

$$\frac{dQ_{gw}}{dt} = -\frac{1}{k}Q_{gw} \tag{6.8}$$

As you may have noticed, this is a simple linear differential equation of first order, and its solution can be found by separating the variable Q_{gw} from the constant and integrating:

$$\int \frac{dQ_{gw}}{Q_{gw}} = -\int \frac{1}{k}dt \tag{6.9}$$

$$\ln(Q_{gw}) = -\frac{t}{k} + c \tag{6.10}$$

$$Q_{gw,t} = e^{-t/k+c} = e^{c}e^{-t/k} \tag{6.11}$$

where $Q_{gw,t}$ is the groundwater outflow rate at time t, and note that c is the constant of integration, and e is also a constant number, so we can replace e^{c} by a new constant, say, C. We obtain,

$$Q_{gw,t} = Ce^{-t/k} \tag{6.12}$$

This equation tells us that if we know the two constants k, which is a liner reservoir constant, and the new constant C, we will know the value of groundwater outflow (Q_{gw}) at any time t. We will talk about the value of k later, but first let us find out the constant C. To do that, we substitute $t=0$ in Eq. (6.12), that is

$$Q_{gw,t=0} = Q_{gw,0} = C \tag{6.13}$$

where $Q_{gw,0}$ is the initial groundwater outflow (at $t=0$), and it shows that the constant C is equal to the value of the groundwater flow at the beginning of the computation. So finally replacing C we get

$$Q_{gw,t} = Q_{gw,0}e^{-t/k} \tag{6.14}$$

This is the linear reservoir method without the recharge term for groundwater outflow simulation. This is also the equation for exponential recession, because it simulates $Q_{gw,t}$ as an exponential decrease of the initial, $Q_{gw,0}$ (because here t and k are both positive values, so $e^{-t/k}$ is always less than unity). This equation also tells us that the groundwater flow value at any time is directly related to the initial value of the flow. In mathematics it is called an initial value problem, meaning that knowing the initial value of a variable its state at any time in the future can be computed. But of course, we need to know the value of the constant k. We know that exponential functions are quite sensitive to the value of the exponent (in this case t/k), so it allows the shape to vary from very sharp decay to nearly flat.

For a step by step computation, Eq. (6.14) can be written as

$$Q_{gw,t} = Q_{gw,t-1}e^{-\Delta t/k} \tag{6.15}$$

where Δt is the computation time step or the time interval between t and $t-1$.

6.4.1.2 Case with recharge

If the recharge is not zero, in other words if the computation needs to be carried out when the recharge is still continuing, or if we run a continuous simulation that includes the periods with and without recharge, we have to use Eq. (6.7) with the recharge term. A standard technique to solve this equation is using a finite difference method as following. Rearranging the linear reservoir Eq. (6.7), we write

$$k\frac{dQ_{gw}}{dt} + Q_{gw} = W_{rcg} \tag{6.16}$$

Using a finite differential approximation, we can write

$$k\frac{Q_{gw,t} - Q_{gw,t-1}}{\Delta t} + \frac{Q_{gw,t} + Q_{gw,t-1}}{2} = W_{rcg} \tag{6.17}$$

Note that W_{rcg} is averaged for the given time step. Finally, using a weighting factor, w, Eq. (6.17) can be written as (see Box 6.1 for the derivation steps)

$$Q_{gw,t} = wW_{rcg} + (1-w)Q_{gw,t-1} \tag{6.18}$$

$$w = \frac{\Delta t}{k + 0.5\Delta t} \tag{6.19}$$

The Eq. (6.18) without recharge (i.e. $W_{rcg} = 0$) would be

$$Q_{gw,t} = (1-w)Q_{gw,t-1} \tag{6.20}$$

Comparing Eq. (6.20) with (6.15), we see that these two equations are identical with

Box 6.1 Finite difference approximation of the linear reservoir

$$k\frac{Q_{gw,t} - Q_{gw,t-1}}{\Delta t} + \frac{Q_{gw,t} + Q_{gw,t-1}}{2} = W_{rcg}$$

$$Q_{gw,t}\left(\frac{k}{\Delta t} + \frac{1}{2}\right) - \left(\frac{k}{\Delta t} - \frac{1}{2}\right)Q_{gw,t-1} = W_{rcg}$$

$$Q_{gw,t}\left(\frac{2k + \Delta t}{2\Delta t}\right) - \left(\frac{2k - \Delta t}{2\Delta t}\right)Q_{gw,t-1} = W_{rcg}$$

$$Q_{gw,t} = \frac{2\Delta t}{2k + \Delta t}W_{rcg} + \frac{2\Delta t}{2k + \Delta t}\frac{2k - \Delta t}{2\Delta t}(Q_{gw,t-1})$$

$$Q_{gw,t} = \frac{\Delta t}{k + 0.5\Delta t}W_{rcg} + \frac{2k + \Delta t - 2\Delta t}{2k + \Delta t}(Q_{gw,t-1})$$

$$Q_{gw,t} = \frac{\Delta t}{k + 0.5\Delta t}W_{rcg} + \left(1 - \frac{\Delta t}{k + 0.5\Delta t}\right)Q_{gw,t-1}$$

$$Q_{gw,t} = wW_{rcg} + (1-w)Q_{gw,t-1}$$

$$e^{-\Delta t/k} = 1 - w = \frac{k - 0.5\Delta t}{k + 0.5\Delta t} \tag{6.21}$$

Thus, we may also write the liner reservoir equation with a recharge (or inflow) with the exponential function as below (Eq. 6.22). This form of the groundwater discharge (or baseflow) equation is also used in the catchment modelling tool SWAT (Neitsch et al., 2009, p. 174).

$$Q_{gw,t} = \left(1 - e^{-\Delta t/k}\right) W_{rcg} + e^{-\Delta t/k} Q_{gw,t-1} \tag{6.22}$$

Because the coefficients in Eqs. (6.18) and (6.22) are in different forms they are not identical. When $k = 2\Delta t$, $e^{-\Delta t/k}$ is nearly equal to 1-w, and so at that ratio ($\Delta t/k = 0.5$), the Q_{gw} calculated from both methods are nearly equal.

6.4.2 Physically based method for groundwater flow modelling

In the conceptual linear reservoir method, we noted that the actual shape of the aquifer is not known. It is simply assumed like a 'reservoir' which can store and release water. One implicit assumption of the shape is the uniform thickness, which is needed to convert the volume per time (e.g. m^3/s) to depth per time (e.g. m/day) and vice versa with the given constant surface area. As a result, spatial variation of water table or heads in the aquifer cannot be estimated with a lumped linear reservoir method. In the physically based model an aquifer is represented as a 3D object, with specified/known length, width and height. Moreover, the water table or piezometric surface are not the same everywhere and can also vary over time. With or without knowing the shape, what is necessary for a hydrological model is to know the amount of groundwater discharge to the river at every time step. In the conceptual method, because the shape as well the water table are unknown, some kind of assumption is necessary to estimate the flow, which in the linear reservoir is the linear proportionality of the storage and outflow. In the physically based method, it is governed by the Darcy's equation, that is the flow rate per unit length is equal to the hydraulic conductivity times the head difference between the river and its surrounding aquifer. There is however one essential principle both conceptual and physically based methods have to maintain, which is the conservation of mass defined by the continuity equation.

So, while the linear reservoir equation for groundwater flow is derived combining the continuity (change in storage = inflow − outflow) and the linear relationship between storage and the outflow, the physically based equation is derived by combining the continuity in 3D with the Darcy's law. Actually, we also discussed the similarly derived equation in Chapter 4 for water flow in unsaturated soil. The 3D groundwater flow equation takes the form

$$\frac{\partial}{\partial x}\left(K_{xx}\frac{\partial h}{\partial x}\right) + \frac{\partial}{\partial y}\left(K_{yy}\frac{\partial h}{\partial y}\right) + \frac{\partial}{\partial z}\left(K_{zz}\frac{\partial h}{\partial z}\right) + W_{ss} = S_s\frac{\partial h}{\partial t} \tag{6.23}$$

where K is the saturated hydraulic conductivity [LT^{-1}], h is hydraulic head [L], S_s is specific storage [L^{-1}] (see Section 6.3), and W_{ss} is the sum of all source or sink terms [$L^3/L^3/T = T^{-1}$], that is the volume of water added or removed per unit volume of the aquifer per unit time. Note that the hydraulic conductivity K in the UZ flow (Chapter 4) is presented as a function of the saturation level (because the degree of saturation can vary both on space and time, and so the K). Groundwater flow is a saturation zone flow, and so the saturated K is used. However, the K_{sat} can vary on different directions of flow (i.e. anisotropic). So, here K is defined separately for the three perpendicular directions of flow (x, y and z): K_{xx}, K_{yy} and K_{zz}.

The term $K_{xx}(\partial h/\partial x)$ from the Darcy's law represents the flow rate q_x in the x-direction, and similarly q_y and q_z in the y and z directions. The left-hand side of the equation takes account of all the volume of water added to or removed from a control volume including the external source (added) or sink (removed), and the right-hand side represents the change in storage in the control volume to balance the net difference in volume added and volume removed on the left-hand side (which is basically the continuity).

As we also discussed in earlier chapters, in practice the partial differential equation of this complexity is solved using a numerical method. For a finite difference numerical solution of a 2D equation, the model domain is usually represented on the x-y plain by rectangular grid cells of lengths Δx and Δy. For the groundwater equation in 3D, we need a grid cell (block) of lengths Δx, Δy and Δz. To represent the continuity in a 3D grid cell in an aquifer, we use Eq. (6.3).

A 3D grid cell has six faces: two of areas $\Delta x \Delta y$, say upper and lower faces, two of areas $\Delta x \Delta z$, say front and rear faces, and two with areas $\Delta y \Delta z$, say left and right faces. So there are six internal inflows (positive) and/or outflows (negatives) through each of the six faces. These six in/out flows can be represented conveniently using the hydraulic conductance (C) as $Q = C\Delta h$, with $C = K\,A/L$, where K is the effective hydraulic conductivity [LT^{-1}], A is the cross-sectional area of the flow [L^2] and L is the length of the flow path [L] (Todd and Mays, 2005).

In Fig. 6.2, the centre cell and six adjacent cells are shown with which the centre cell exchanges flows. Thus, the flows entering (+) or leaving (−) through the six faces can be expressed as (see, also in Todd and Mays, 2005):

$$\left.\begin{aligned} Q_{1-0} &= C_{1-0}\left(h_{i-1,j,k} - h_{i,j,k}\right) \\ Q_{2-0} &= C_{2-0}\left(h_{i+1,j,k} - h_{i,j,k}\right) \\ Q_{3-0} &= C_{3-0}\left(h_{i,j-1,k} - h_{i,j,k}\right) \\ Q_{4-0} &= C_{4-0}\left(h_{i,j+1,k} - h_{i,j,k}\right) \\ Q_{5-0} &= C_{5-0}\left(h_{i,j,k-1} - h_{i,j,k}\right) \\ Q_{6-0} &= C_{6-0}\left(h_{i,j,k+1} - h_{i,j,k}\right) \end{aligned}\right\} \tag{6.24}$$

The Q's and C's in the above equations are between the adjacent cells. For example, Q_{1-0} is the flow between cell 1 and 0, and C_{1-0} is the conductance between cell 1 and 0. The flow can be in either direction depending on the relative magnitudes of h in the two cells. With the way two h values are represented in the above equations, a positive Q_{1-0} means the flow from cell 1 to cell zero and a negative Q_{1-0} means flow

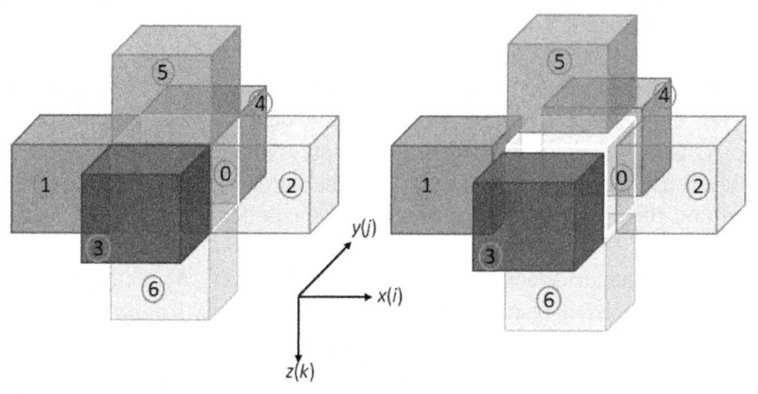

FIG. 6.2

A grid cell (in 3D), numbered 0 with its centre (*i*, *j*, *k*) and six adjacent cells numbered 1 to 6 with which the centre cell exchanges flows. Taking *i* as horizontal *x*-axis, *j* as horizontal *y*-axis and *k* as vertical *z*-axis, the centre coordinates of the six blocks are 1(*i*−1, *j*, *k*), 2(*i*+1, *j*, *k*), 3(*i*, *j*−1, *k*), 4(*i*, *j*+1, *k*), 5(*i*, *j*, *k*−1) and 6(*i*, *j*, *k*+1). The figure in the right is same as in the left, with each cell separated slightly from cell 0 to show the cell more clearly.

from the centre cell 0 to cell 1. Understanding of the C's here requires some elaboration, which we will do in a moment. Let's first write the full equation by substituting these flow terms in Eq. (6.3) and also expanding the right-hand side of the equation to represent the change in h in the centre cell with respect to change in time. The full equation becomes

$$C_{1-0}\left(h_{i-1,j,k}-h_{i,j,k}\right)+C_{2-0}\left(h_{i+1,j,k}-h_{i,j,k}\right)+C_{3-0}\left(h_{i,j-1,k}-h_{i,j,k}\right)+C_{4-0}\left(h_{i,j+1,k}-h_{i,j,k}\right)$$
$$+C_{5-0}\left(h_{i,j,k-1}-h_{i,j,k}\right)+C_{6-0}\left(h_{i,j,k+1}-h_{i,j,k}\right)+W_{ss,(i,j,k)}=S_{s,(i,j,k)}\Delta V_{i,j,k}\frac{h^{n+1}_{i,j,k}-h^{n}_{i,j,k}}{\Delta t} \quad (6.25)$$

One thing that is not indicated in the above equation is the time indication of the piezometric heads on the left-hand side of the equation. On the right-hand side, we specified the time level by n and $n+1$. The time level n refers to the current time level in the simulation, which is usually the time step all the state variables, such as h, are known and $n+1$ refers to one-time step ahead of the current time step, and so values of the state variables are unknown at $n+1$ time step. So, an easier solution would be to take all the heads on the left-hand side of the equation from the current time step, n. The problem however is that it does not work well in practice. Note that one cell has to exchange to six neighbouring cells, except at the boundaries or cells with a constant head. In terms of the numerical scheme, such a solution technique is called an explicit scheme, which quickly becomes unstable before the solution is found. A fully implicit scheme, e.g. the one also used by McDonald and Harbaugh (1984) and in the popular groundwater flow model MODFLOW (Harbaugh et al., 2000), is to use all the h values from the unknown time step, $n+1$. This inevitably requires iterations and the solutions for all the cells have to be sought simultaneously at each time step. Thus, finally, a finite difference implicit groundwater flow 3D equation is

$$C_{1-0}\left(h_{i-1,j,k}^{n+1}-h_{i,j,k}^{n+1}\right)+C_{2-0}\left(h_{i+1,j,k}^{n+1}-h_{i,j,k}^{n+1}\right)+C_{3-0}\left(h_{i,j-1,k}^{n+1}-h_{i,j,k}^{n+1}\right)+C_{4-0}\left(h_{i,j+1,k}^{n+1}-h_{i,j,k}^{n+1}\right)$$
$$+C_{5-0}\left(h_{i,j,k-1}^{n+1}-h_{i,j,k}^{n+1}\right)+C_{6-0}\left(h_{i,j,k+1}^{n+1}-h_{i,j,k}^{n+1}\right)+W_{ss,(i,j,k)}=S_{s,(i,j,k)}\Delta V_{i,j,k}\frac{h_{i,j,k}^{n+1}-h_{i,j,k}^{n}}{\Delta t} \tag{6.26}$$

Most of the details presented under the physically based method are similar to the ones used in 'specialist' groundwater modelling tools, such as MODFLOW. A detailed physically based approach for groundwater modelling is rarely used in a catchment hydrological model as an integral component. Among the 17 catchment hydrological models reviewed in this chapter (Section 6.5), only MIKE SHE (DHI, 2017a) has the physically based 3D method (as an option) for groundwater flow. Most models have some kind of linear reservoir type (conceptual) methods with varied levels of details.

6.4.2.1 Horizontal and vertical conductance between two adjacent cells

The conductance C_{1-0}, for example, is the conductance between cells 1 and 0, which means C_{1-0} depends on the hydraulic conductivity and dimensions of the two cells adjacent to each other. Let us first use a simple case for the conductance, which is if

(1) The horizontal and vertical hydraulic conductivities K_{xx}, K_{yy} and K_{zz} are constant for all cells, and
(2) Cell dimensions Δx, Δy and Δz are constant for all rows and columns.

In this case the six conductivities used in Eqs. (6.25) and (6.26) become

$$\left.\begin{array}{l} C_{1-0}=C_{2-0}=K_{xx}\Delta y\Delta z/\Delta x \\ C_{3-0}=C_{4-0}=K_{yy}\Delta x\Delta z/\Delta y \\ C_{5-0}=C_{6-0}=K_{zz}\Delta x\Delta y/\Delta z \end{array}\right\} \tag{6.27}$$

If the hydraulic conductivities and cell dimensions are not constant as in the simplest case described above, we can derive an equivalent conductance between two cells from the conductance values of one half of each cell. To work out a general equation, let us assume M and N are two adjacent cells and the grid points where the heads are defined are at the centre of each cell. One way to estimate the equivalent conductance makes two assumptions. First, the flow through each half of cell M, say $Q_{\text{m/2}}$, and N, say $Q_{\text{n/2}}$, are equal, and for continuity both are equal to the total flow between the cells; that is $Q_{\text{m-n}}=Q_{\text{m/2}}=Q_{\text{n/2}}$. Second, the head difference between the two centres, say $\Delta h_{\text{m-n}}$, is the sum of the head difference between each centre and the boundary between the two, say $\Delta h_{\text{m/2}}$ and $\Delta h_{\text{n/2}}$, that is $\Delta h_{\text{m-n}}=\Delta h_{\text{m/2}}+\Delta h_{\text{n/2}}$. Similarly, say $C_{\text{m/2}}$ and $C_{\text{n/2}}$ are the conductance values of the two half cells and $C_{\text{m-n}}$ is the equivalent conductance between the cells, we can write

$$\left.\begin{array}{l} \Delta h_{m-n} = \Delta h_{m/2} + \Delta h_{n/2} \\ \dfrac{Q_{m-n}}{C_{m-n}} = \dfrac{Q_{m/2}}{C_{m/2}} + \dfrac{Q_{n/2}}{C_{n/2}} \\ \dfrac{1}{C_{m-n}} = \dfrac{1}{C_{m/2}} + \dfrac{1}{C_{n/2}} \\ C_{m-n} = \dfrac{C_{m/2}C_{n/2}}{C_{m/2}+C_{n/2}} \end{array}\right\} \tag{6.28}$$

Using the relationship for the equivalent conductance (Eq. 6.28), the conductance for the boundary between the cells can be estimated, which is same as using the harmonic mean of the conductance values of the two adjacent cells. Of the six conductance terms we need, four are horizontal conductance (in the x- and y-axes) and two vertical conductance (in the z-axis). The four horizontal conductance terms are shown below with the following conditions: horizontal and vertical conductivities can be different in different cells, Δx is contact along the given row in y-axis (that is $\Delta x_{i,j_{-1},k} = \Delta x_{i,j,k} = \Delta x_{i,j+1,k} = \Delta x$), and Δy is constant the given x-axis row (that is $\Delta y_{i-1,j,k} = \Delta y_{i,j,k} = \Delta y_{i+1,j,k} = \Delta y$). Recalling that conductance $C = KA/L$, we can express C_{1-0} in terms of K as following:

$$\left.\begin{array}{l} \dfrac{1}{C_{1-0}} = \dfrac{1}{\dfrac{K_{i-1,j,k}\Delta y \Delta z_{i-1,j,k}}{(\Delta x_{i-1,j,k})/2}} + \dfrac{1}{\dfrac{K_{i,j,k}\Delta y \Delta z_{i,j,k}}{(\Delta x_{i,j,k})/2}} \\ C_{1-0} = \dfrac{2(\Delta y \Delta z_{i-1,j,k}\Delta z_{i,j,k})K_{(i-1,j,k)}K_{i,j,k}}{\Delta x_{i,j,k}\Delta z_{i-1,j,k}K_{i-1,j,k} + \Delta x_{i-1,j,k}\Delta z_{i,j,k}K_{i,j,k}} \end{array}\right\} \tag{6.29}$$

Similarly, the other three are

$$\left.\begin{array}{l} C_{2-0} = \dfrac{2(\Delta y \Delta z_{i,j,k}\Delta z_{i+1,j,k})K_{i,j,k}K_{i+1,j,k}}{\Delta x_{i+1,j,k}\Delta z_{i,j,k}K_{i,j,k} + \Delta x_{i,j,k}\Delta z_{i+1,j,k}K_{i+1,j,k}} \\ C_{3-0} = \dfrac{2(\Delta x \Delta z_{i,j-1,k}\Delta z_{i,j,k})K_{i,j-1,k}K_{i,j,k}}{\Delta y_{i,j,k}\Delta z_{i,j-1,k}K_{i,j-1,k} + \Delta y_{i,j-1,k}\Delta z_{i,j,k}K_{i,j,k}} \\ C_{4-0} = \dfrac{2(\Delta x \Delta z_{i,j,k}\Delta z_{i,j+1,k})K_{i,j,k}K_{i,j+1,k}}{\Delta y_{i,j+1,k}\Delta z_{i,j,k}K_{i,j,k} + \Delta y_{i,j,k}\Delta z_{i,j+1,k}K_{i,j+1,k}} \end{array}\right\} \tag{6.30}$$

For the two vertical conductance terms, first for C_{5-0}, we can write

$$\left.\begin{array}{l} \dfrac{1}{C_{5-0}} = \dfrac{1}{\dfrac{K_{i,j,k-1}\Delta x \Delta y}{(\Delta z_{i,j,k-1}/2)}} + \dfrac{1}{\dfrac{K_{i,j,k}\Delta x \Delta y}{(\Delta z_{i,j,k}/2)}} \\ C_{5-0} = \dfrac{\Delta x \Delta y}{\dfrac{(\Delta z_{i,j,k-1}/2)}{K_{i,j,k-1}} + \dfrac{(\Delta z_{i,j,k}/2)}{K_{i,j,k}}} \end{array}\right\} \tag{6.31}$$

Finally, the C_{6-0} is

$$C_{6-0} = \frac{\Delta x \Delta y}{\dfrac{(\Delta z_{i,j,k}/2)}{K_{i,j,k}} + \dfrac{(\Delta z_{i,j,k+1}/2)}{K_{i,j,k+1}}} \tag{6.32}$$

Note that in Eqs. (6.29) and (6.30), the hydraulic conductivities are K_{xx} for C_{1-0} and C_{2-0}, and K_{yy} for C_{3-0} and C_{4-0}. Similarly, in Eqs. (6.31) and (6.32), that is for C_{5-0} and C_{6-0}, the hydraulic conductivities are K_{zz}. In these equations, the subscripts xx, yy and zz for K are omitted for conciseness. The equations for horizontal conductance represented by Eqs. (6.29) and (6.30) can be conveniently expressed using transmissivity, i.e. by substituting $T_{xx} = K_{xx}\Delta z$ and $T_{yy} = K_{yy}\Delta z$, where T_{xx} and T_{yy} are the transmissivities (dimension $L^2 T^{-1}$) in the x-axis and y-axis directions, respectively (McDonald and Harbaugh, 1984). For example, for C_{1-0} and C_{2-0} are shown here (Eq. 6.33).

$$\left.\begin{aligned} C_{1-0} &= \frac{2\Delta y T_{xx(i-1,j,k)} T_{xx(i,j,k)}}{\Delta x_{i,j,k} T_{xx(i-1,j,k)} + \Delta x_{i-1,j,k} T_{xx(i,j,k)}} \\ C_{2-0} &= \frac{2\Delta y T_{xx(i,j,k)} T_{xx(i+1,j,k)}}{\Delta x_{i+1,j,k} T_{xx(i,j,k)} + \Delta x_{i,j,k} T_{xx(i+1,j,k)}} \end{aligned}\right\} \tag{6.33}$$

More cases of conductance formulations can be derived based on the modelling dimensionality (1D, 2D or 3D) used, variation of hydraulic conductivity (anisotropic or isotropic), variation of cell sizes (Δx, Δy and Δz) and type of aquifers (confined or unconfined) (see, e.g. McDonald and Harbaugh, 1984; Todd and Mays, 2005; DHI, 2017a).

6.4.2.2 Differences between confined and unconfined aquifers

In the application of groundwater flow equations, differences between confined and unconfined aquifers need to be taken care of. In an unconfined aquifer the hydraulic head is the same as the water table, and as water table changes the saturated thickness of the aquifer changes. This has implications in the estimation of the conductance during simulation. Another difference which is discussed in Section 6.3 is related to the mechanism with which volume changes are adjusted in confined and unconfined aquifers. Because of this the specific storage (S_s) is used for a confined aquifer and specific yield (S_y) for an unconfined aquifer. In a groundwater system consisting of confined/unconfined aquifer grid cells, a confined grid cell may become unconfined during simulation. A general approach to handle this is by allowing transition between confined-unconfined situation in any grid cell. More details on this can be found in MODFLOW and MIKE SHE (saturated zone) technical manuals.

6.4.2.3 Sources and sinks

As we described earlier, the first six terms of Eq. (6.25) take care of the flow movement from one grid cell in the model domain to another. But it may be that one or more of the grid cells are receiving flow from a source external to the aquifer grid cells or flow moving out from a cell to a location external to the aquifer grid cells. These are the sources (from where the flow is coming in) and sinks (to where the flow is going out). For example, the recharge from the rainwater through the overlaying soil layer is a common source term. In a catchment with aquifer–river connections, the river can be both a source or a sink depending on the hydraulic heads in the river and the aquifer grid cell it is connected to. Other examples include pumping from a well a sink, and injection well a source in a pump and treat experiment (see, e.g., Maskey et al., 2002). A connection to a large surface water body (lake or reservoir) can be both a source or sink. In the groundwater flow Eqs. (6.23)–(6.25), the source/sink term W_{ss} is taken positive (+) if it is a source and negative (−) if a sink. Note that whether the term should be a positive or negative is sometimes confusing because it depends whether the term is on the left-hand side or right-hand side of the equation. In the above equations, if the term is put on the right-hand side then a source becomes negative (−) and a sink a positive (+). The best way to avoid this confusion is to put the source/sink terms always on the left-hand side of the equation so the same convention as in the water balance equation applies.

In general, the source/sink terms can be represented by two functions, say f and g, with f dependent on the head in the cell and g independent of the head in the cell, or a constant (McDonald and Harbaugh, 1984). That is

$$W_{ss(i,j,k)} = f\left(h_{i,j,k}\right) + g \tag{6.34}$$

For example, the recharge from rainfall (a source) is independent of the head in the cell, whereas discharge to or recharge from a river is dependent to the head in the aquifer cell.

6.4.2.4 Initial and boundary conditions

To simulate groundwater flow, we need to write the groundwater continuity Eq. (6.25) or (6.26) for every grid cell in the model domain whose head can vary (i.e. variable head cell) during the simulation. For each of these cells for which the equation needs to be solved, initial values of heads (h) called initial condition must be specified. These are the heads at the starting time of the simulation, i.e. $h\ (i, j, k, t_0)$.

In addition to the initial condition, boundary conditions are required for all the cells that are at the boundaries of the model domain. Boundary conditions define the conditions of the model variables (e.g. water head and flow) at the boundary of the model domain. While the initial condition value is for time $t = t_0$ for all cells, the boundary condition is required for entire time of simulation for all boundary cells.

The boundary conditions can be classified as no-flow boundary, specified head boundary, specified flux boundary and head dependant flux boundary. A no-flow boundary is effectively an inactive cell or impervious cell with no-flow in and out from it. Which means, cells adjacent to a 'no-flow' boundary cell neither receive from nor lose water to the boundary. An easy way to implement this condition in the model is to specify zero conductance for the no-flow boundary cell. The lower boundary of the groundwater model is always taken as the impermeable basement of the aquifer so it is a no-flow boundary.

In the specified head boundary, the head (h) in the boundary cell is specified for the entire simulation period, which can be constant or transient. If the specified head boundary is a constant value, it is also called a constant head boundary. A constant head boundary is usually defined when the grid cell is in contact with a large water body, such as lake or river.

In the specified flux boundary, the flow rate (Q) is specified, as opposed to heads, for the entire period of simulation. Lateral inflows from or outflows to a neighbouring catchment are specified as specified flux boundaries. No-flow boundary is also a special case of specified flux boundary (zero flux).

In the head dependent flux boundary, the flow to or from the boundary is dependent to the head difference. A river cell connected to the aquifer is an example of a head dependent flux boundary when the river is chosen as the model boundary. In which case the groundwater discharge to a river cell or recharge from a river cell to the aquifer is dependent on the head difference between the river cell and aquifer cell.

6.4.2.5 Steady-state and transient simulation

In groundwater flow modelling, we often hear about steady-state and transient flow simulations. But what is really the difference between the two? When is a groundwater flow system in a steady-state and when transient in practice?

In terms of the flow equation, the answer to the first question is quite simple. The steady-state assumes that the storage change in every grid cell at any time is zero. This means that the right-hand side of Eq. (6.25) is zero, so the resulting equation is simply

$$\begin{aligned} &C_{1-0}\left(h_{i-1,j,k}-h_{i,j,k}\right)+C_{2-0}\left(h_{i+1,j,k}-h_{i,j,k}\right)+C_{3-0}\left(h_{i,j-1,k}-h_{i,j,k}\right)+C_{4-0}\left(h_{i,j+1,k}-h_{i,j,k}\right) \\ &\quad +C_{5-0}\left(h_{i,j,k-1}-h_{i,j,k}\right)+C_{6-0}\left(h_{i,j,k+1}-h_{i,j,k}\right)+W_{ss,(i,j,k)}=0 \end{aligned} \tag{6.35}$$

Note that Eq. (6.25) is used here instead (6.26) because the time index $n+1$ used for h in Eq. (6.26) is irrelevant for the steady-state simulation.

But what does this mean in reality? The steady-state flow Eq. (6.35) says that the summation of all the flows in and out of a cell including sources and sinks equals to zero. This means that the sum of the inflows (including sources) and sum of the outflows (including sinks) are equal in every cell. So in practice a steady-state situation exists if there is a constant inflows for a long enough period so as to bring the system in an equilibrium between the inflows and outflows effectively making no change in storage in the system. Once it is in a steady-state equilibrium, it will be maintained as long as the sources and sinks remain constant. At this point, from the modelling point of view, a question may arise: what does the model do in the steady-state simulation? Actually, in the steady-state simulation, the model looks for redistributing the groundwater heads in every cell such that the equilibrium between the inflows and outflows is obtained. Thus, in the steady-state, the heads in different cells can differ from each other (heads are spatially varying), but the head in each cell remains constant over time.

So, the groundwater heads established from a steady-state model is valid as long as there are no changes in the input or output, or more specifically, no changes in sources or sinks and boundary flow or heads. However, very often we find many inflows/outflows that do constantly change. For example, the recharge changes depending on rainfall and UZ soil moisture condition. If there are pumping wells, they will not pump at the same rate all the time. If the aquifer is connected to a river, water level in the river does change and so does the aquifer discharge to the river, etc. So basically with the steady-state model we simulate the groundwater heads with respect to average values of the fluxes (e.g. recharge, discharge to the river, pumping from well, etc.) over a certain period of time.

Let us simplify the steady-state equation further. First, assume a 2D model with no source and sink terms. Then Eq. (6.35) becomes

$$C_{1-0}\left(h_{i-1,j}-h_{i,j}\right)+C_{2-0}\left(h_{i+1,j}-h_{i,j}\right)+C_{3-0}\left(h_{i,j-1}-h_{i,j}\right)+C_{4-0}\left(h_{i,j+1}-h_{i,j}\right)=0 \quad (6.36)$$

Second, assume a homogenous and isotropic aquifer of uniform thickness. These conditions imply that $C_{1-0}=C_{2-0}=C_{3-0}=C_{4-0}=C$ and result in

$$h_{i,j}=\frac{h_{i-1,j}+h_{i+1,j}+h_{i,j-1}+h_{i,j+1}}{4} \quad (6.37)$$

This equation tells that the hydraulic head at any grid cell—other than those specified as boundary cells—is equal to the average of the heads of the four cells adjacent to it. If we write this equation for all the unknown cells, we will have as many numbers of liner algebraic equations as number of grid cells with unknown heads. Such a set of algebraic equation can be solved conveniently using an iterative method such as Guess-Seidel. The algorithm for Guess-Seidel method can be found in Bear and Verruijt (1987) and also in most standard engineering mathematics text books, e.g. Kreyszig (1983), Duchateau (1992) and Harris and Stocker (2006).

Note that a steady-state solution of a groundwater model is not particularly useful for integration with a catchment hydrological model that runs on a daily or smaller time step. From a catchment modelling perspective, the primary function of a groundwater model is to simulate the baseflow (or groundwater flow) to the river, which is most often a transient discharge.

6.5 How different catchment hydrological models treat groundwater (saturated zone) flow?

Brief descriptions of methods and approaches used in 17 hydrological models for groundwater (saturated zone) flow are presented here (Table 6.1). The selected models are among the widely used models, but the list is not exhaustive, and the purpose of presenting the table is not for giving the author's judgement about the models. It is simply intended to describe the key concepts of the methods and approaches used in the models. The descriptions attempt to capture the main essence or features, but may not be detailed enough to provide complete details of the methods, for which the referenced literature should be consulted.

Table 6.1 Groundwater (saturated zone) flow methods used in various hydrological models. (Abbreviations used in the table include GW: groundwater, LR: linear reservoir, SZ: saturated zone, and UZ: unsaturated zone.)

Modelling software	Method used	Additional information
BROOK90	It allows more than one soil layers for the UZ. Recharge to GW is calculated from the water in the bottommost soil layer as a constant fraction (0 to 1) of the hydraulic conductivity of the layer. From the GW storage (S), groundwater flow to the river (Q_{gw}) and seepage loss (Seep) are calculated as constant fractions of the storage, which are controlled by two coefficients (Federer, 2002). Say, the two coefficients controlling Q_{gw} and Seep are f_1 and f_2, and total release from groundwater at each time step is Rel, then $$\mathrm{Rel} = f_1 S$$ $$\mathrm{Seep} = f_1 f_2 S$$ $$Q_{gw} = \mathrm{Rel} - \mathrm{Seep} = f_1(1-f_2)S$$	
CASC2D	It is primarily an infiltration-excess (Hortonian) surface flow simulation model, and it does not have a GW component or baseflow as such; instead the soil layer (UZ) is assumed infinitely deep (Ogden, 1998; Downer et al., 2002).	
CHARM (also WATFLOOD)	Subsurface is divided into three layers: the upper zone (UpZ), intermediate zone (IZ) and lower zone (LZ). The GW (baseflow) is taken from the LZ. Recharge to the LZ is a simple fraction (called an IZ resistance) of moisture storage minus moisture retention in the UpZ. The GW discharge is a simple power function of the LZ storage, with two parameters: a multiplying factor and a power to the LZ storage. The interflow and GW flow are estimated simultaneously (Kouwen, 2018).	CHARM: Canadian Hydrological and Routing Model
Flo2D	It does not have a GW component and so no groundwater discharge. It is primarily an event-based overland flow model for flood events, and so essentially not meant for continuous (long-term) simulation of rainfall-runoff. The infiltration is considered as a loss term (FLO-2D, 2003).	
HBV	It has one soil layer (UZ) and two groundwater layers (SZ). Precipitation and snowmelt are divided into the soil layer and GW layer using a power function of the current soil moisture ratio to the maximum capacity. If the UZ is holding more water, more water (from precipitation + soil moisture) becomes recharge to the upper SZ. Water from the upper GW layer to the	

Continued

Table 6.1 Groundwater (saturated zone) flow methods used in various hydrological models. (Abbreviations used in the table include GW: groundwater, LR: linear reservoir, SZ: saturated zone, and UZ: unsaturated zone.)—cont'd

Modelling software	Method used	Additional information
	lower layer is based on a specified percolation rate. Groundwater flow to the river is taken from the two GW layers as a direct fraction (recession coefficient) of the storage in each layer. In case the storage in the upper layer exceeds a specified threshold, a third flow component is added as a fraction of the excess storage value. The sum of the two (or three) flow components is transformed (or smoothened) using a "triangular weighting function" (Bergström, 1992; Seibert, 2005).	
HEC-HMS	It has the LR method with the Soil Moisture Accounting (SMA) model and exponential recession method (Feldman, 2000). It also allows a user defined monthly constant baseflow input. In the LR with SMA, two GW layers are possible, which are analogous to shallow and deep aquifers. There is also a possibility for percolation from the second layer GW which is a loss in the catchment water balance. The recharge rates to the upper GW layer from the soil storage and to the second GW layer from the first layer are estimated based on the specified maximum percolation rate and relative moisture level (i.e. existing/maximum) in the two layers. The exponential recession is meant for event simulations.	
LISFLOOD	It has two groundwater layers (upper and lower). Similar to the HBV model, the outflow is taken as a constant liner fraction of the storage. The outflow from both the upper and lower GW layers are added to the nearest stream cell (note that LISFLOOD is a grid based distributed model) within the same time step without any delay function (Burek et al., 2013). Recharge from the soil layer to the upper GW layer is based on two processes: recharge of excess moisture after saturation based on a function for variable infiltration capacity and preferential bypass flow. The recharge from the upper GW layer to the lower is based on a specified percolation rate as much as that is available in the upper GW layer. There is also a possibility to allow deep percolation loss from the lower GW layer.	
MIKE SHE	It has two options: the physically based 3D method (that is the 3D continuity equation with Darcy's equation) and the LR method (DHI, 2017a).	

Table 6.1 Groundwater (saturated zone) flow methods used in various hydrological models. (Abbreviations used in the table include GW: groundwater, LR: linear reservoir, SZ: saturated zone, and UZ: unsaturated zone.)—cont'd

Modelling software	Method used	Additional information
	The physically based method can be run as steady-state or transient, and three types of boundaries (at the boundary grid cells) are allowed: constant or time-varying head, head gradient, and head dependent flux). Sources (e.g. injection) and sinks (pumping) can be applied to any cells / any layers. GW flow to the river (connected to any grid cell) is a sink term and calculated using the Darcy's equation (that is conductance times head gradient). River cells may also drain to an aquifer cell (a source term) when the GW table is below the river water level. Recharge to an unconfined aquifer is calculated in the UZ model. If the UZ is not included in the model, it the recharge values can be directly specified. The LR method is applied to subcatchments (as opposed to grid cells in the physically based method). Each subcatchment will have at least two layers of reservoirs: the upper layer reservoir (also called interflow layer) and lower or deeper layer (also called a baseflow layer). There can be more than one reservoirs (assumed at different levels) in the interflow layer and more than one deeper (baseflow) layers. An interflow reservoir drains to the baseflow reservoir and to another interflow reservoir at the lower level, or to the river in the subcatchment. The baseflow reservoirs drain to the river in the same subcatchment. Inflow to the interflow reservoir is infiltration from the UZ and to the baseflow reservoir is from the interflow reservoirs.	
NAM	It has three storages (reservoirs): one Surface Storage (SS), and two subsurface storages—the Root Zone (RZ), also called a lower zone, and GW (DHI, 2017b). A portion of the excess water from the SS after surface runoff is added (as recharge) to the GW which is controlled by relative moisture storage in the RZ (that is, the ratio of the current moisture content to max moisture content) and a threshold RZ moisture content to allow recharge to GW. Water from the GW storage is routed as baseflow with a liner reservoir time constant. A second (or lower) GW layer can also be specified, in which case a fraction of the recharge to the first GW layer will be passed to the second layer.	

Continued

Table 6.1 Groundwater (saturated zone) flow methods used in various hydrological models. (Abbreviations used in the table include GW: groundwater, LR: linear reservoir, SZ: saturated zone, and UZ: unsaturated zone.)—cont'd

Modelling software	Method used	Additional information
PCR-GLOBWB	It has two soil layers and underneath a GW storage layer. The first (top) soil layer controls surface runoff and infiltration to the second layer. The infiltration from layer 1 to 2 is calculated as unsaturated hydraulic conductivity (K) of layer 1, and from layer 2 to GW storage (third layer) as unsaturated K of layer 2 (Van Beek and Bierkens, 2008). The baseflow is taken as a fraction of the current GW storage defined by a reservoir coefficient. The reservoir coefficient represents the average residence time of the aquifer, and unlike most other models where this parameter is usually determined by calibration, here it is estimated using aquifer properties (in particular the saturated K, specific yield, aquifer depth and drainage length) with a lumped schematization of aquifer in the catchment.	
PRMS (Ver. 4) also used in GSFLOW	The gravity flow reservoir in the subsurface soil layer drains to the groundwater reservoir. Outflows from the GW reservoir are baseflow and a sink, which is a loss from the catchment water balance. Both are estimated as a liner fraction of the GW storage. The baseflow generated from a reservoir can be routed to a downstream GW reservoir or to a stream based on a defined connections between GW reservoirs and reservoirs and streams (Markstrom et al., 2015; Markstrom et al., 2008). In GSFLOW, selected modules (computational components) from PRMS and MODFLOW are coupled to allow more comprehensive modelling of surface water—groundwater integration.	PRMS: Precipitation Runoff Modeling System (Markstrom et al., 2015). GSFLOW: Coupled Ground-water and Surface-water Flow Model (Markstrom et al., 2008).
RRI model	It does not have groundwater (or baseflow) as such, but the subsurface flow is lumped into a single later subsurface flow. Both the surface flow and lateral subsurface flow in the non-river grid cells are routed using 2D equations of mass and momentum balance (Sayama, 2017).	
SWAT	It has two groundwater layers: shallow and deep aquifers. Baseflow is taken from the shallow aquifer using LR method. To estimate the recharge for any time step, first it estimates the amount of water that is ready to leave the unsaturated zone, and applies an exponential decay method to add part of the available water to the aquifers accounting for the	

Table 6.1 Groundwater (saturated zone) flow methods used in various hydrological models. (Abbreviations used in the table include GW: groundwater, LR: linear reservoir, SZ: saturated zone, and UZ: unsaturated zone.)—cont'd

Modelling software	Method used	Additional information
	delay between when the water becomes available and when it reaches the aquifer. This is similar to applying the LR method between unsaturated zone and aquifer. Further, the total recharge for a day from the unsaturated zone (soil layer) is divided into shallow aquifer and deep aquifer using a coefficient called an 'aquifer percolation coefficient'. Deep aquifer is not connected to the river in the catchment, which means recharge to the deep aquifer is a loss, but stored water in the deep aquifer can be pumped out (Neitsch et al., 2009). It allows a limited amount of the water in the shallow aquifer to be taken for evaporation. This amount is limited to a fraction of the evaporation demand (expressed by potential evapotranspiration) of the day and available water in the aquifer above a threshold value.	
UBC model	It has primarily three runoff components; the first is the surface runoff (or fast runoff) and the other two are subsurface runoff components, called the interflow and groundwater (or slow runoff). The runoff volume partition is based on the soil moisture deficit level and a fixed order of priority, i.e. first the surface runoff, then evaporation, then groundwater recharge and finally interflow (middle) component (Quick and Pipes, 1977; Quick, 1995). The recharge to the GW is further divided into upper and deep GW zones. The baseflow is taken from the two GW storages as a fixed percentage of each storage and routed using the LR method.	UBC: University of British Colombia
VIC model	With three-layer soil profiles, the bottom layer (for groundwater baseflow) receives recharge from the upper (second) soil layer at the rate of unsaturated hydraulic conductivity, which is based on the empirical equation defined by Brooks and Corey (1964). The baseflow from the bottom layer is taken using a linear or nonlinear relationship of the storage in the layer (or the moisture content), depending on whether the storage is below or above a defined fraction of the maximum storage of the layer. The routing is based on a linear transfer function approach (Liang et al., 1994; Gao et al., 2010; Lohmann et al., 1998).	VIC: Variable Infiltration Capacity. See also the VIC version 5 webpage: https://vic.readthedocs.io/en/master/

Continued

Table 6.1 Groundwater (saturated zone) flow methods used in various hydrological models. (Abbreviations used in the table include GW: groundwater, LR: linear reservoir, SZ: saturated zone, and UZ: unsaturated zone.)—cont'd

Modelling software	Method used	Additional information
WaSiM	It has two approaches to model surface/subsurface flow: TOPMODEL and the Richards equation. In the TOPMODEL approach, baseflow is calculated for a subcatchment (as opposed to each grid cell) using an exponential function of the saturation deficit and topographic index of the subcatchment. Larger the saturation deficit, less is the baseflow, and smaller the topographic index larger the baseflow. The topographic index is defined as a logarithmic function of catchment and soil characteristics, e.g. the topographic slope and hydraulic transmissivity. As the baseflow is estimated for a subcatchment, no routing is applied to the baseflow. In the Richards equation model, the GW can be modelled by physically based (2D) approach with continuity and Darcy's equations. Baseflow as groundwater discharge to the river cell is taken from the cell/layer connected to the river cell estimated from the Darcy's equation with a restriction for the maximum allowable discharge from a cell based on a defined limiting moisture content in the cell. Recharge to the groundwater (SZ) layer is taken from available moisture at the bottommost layer of the UZ (Schulla, 2012).	See also the version updates on the WaSiM webpage: http://www.wasim.ch/en/the_model/dev_details.htm
Xinanjiang model	It assumes three soil layers (upper, lower and deep) but first total runoff is generated from a catchment or grid cell (depending on a lumped or distributed version) without differentiating the layers. A parameter "tension water storage" (which is defined as the soil moisture at field capacity minus soil moisture at permanent wilting point) is used to estimate the runoff generation. Then it is partitioned into three runoff components: surface flow, interflow and baseflow. Another parameter "free water storage" is used to first estimate the surface runoff (independent of but by definition less than the total runoff), which is then deducted from the total runoff to add to the current free water storage. The free water storage (after the residual of the total runoff is added) is divided into interflow and baseflow using a defined factor (parameter) for each flow component. Note that this applies only to pervious area of the catchment. So there is no explicit estimate of recharge to the deep soil layer. The baseflow is further routed to the catchment outlet through the LR method (Zhao, 1992; Yao et al., 2012; Fang et al., 2017).	

References

Bouwer, H., 1978. Groundwater Hydrology. McGraw-Hill, New York.
Harbaugh, A.W., Banta, E.R., Hill, M.C., McDonald, M.G., 2000. MODFLOW-2000—User Guide to Modularization Concepts and the Ground-Water Flow Process. Open-File Report 00–92, U.S. Geological Survey, Reston, Virginia.
Neitsch, et al., 2009. SWAT technical documentation. Ver 2009.
Bear, J., 1972. Dynamics of Fluids in Porous Media. Dover Publications, Inc., New York.
McDonald, M.G., Harbaugh, A.W., 1984. A modular three-dimensional finite-difference ground-water flow model. U.S. Geological Survey Open-File Report 83-875.
Nonner, J.C., 2015. Introduction to Hydrogeology, third ed. Taylor & Francis Group, London.
Todd, D.K., Mays, L.W., 2005. Groundwater Hydrology, third ed. Wiley.
DHI, 2017a. MIKE SHE Volume 2: Reference Guide. DHI, Denmark. https://manuals.mikepoweredbydhi.help/2017/Water_Resources/MIKE_SHE_Printed_V2.pdf.
Maskey, S., Jonoski, A., Solomatine, D.P., 2002. Groundwater remediation strategy using global optimization algorithms. J. Water Resour. Plann. Manage., ASCE 128 (6), 431–440.
Bear, J., Verruijt, A., 1987. Modelling Groundwater Flow and Pollution. D. Reidel Publishing Company, Dordrecht.
Kreyszig, E., 1983. Advanced Engineering Mathematics, fifth ed. Wiley.
Duchateau, P., 1992. Advanced Mathematics for Engineers and Scientists. Dover Publications, Inc., New York.
Harris, J.W., Stocker, H., 2006. Handbook of Mathematics and Computational Science. Springer, New York.
Federer, C.A., 2002. BROOK 90: a simulation model for evaporation, soil water, and streamflow. http://www.ecoshift.net/brook/brook90.htm.
Ogden, F.L. (1998, rev 2001). A Brief Description of the Hydrologic Model CASC2D. Univ. Connecticut.
Downer, C.W., Ogden, F.L., Martin, W.D., Harmon, R.S., 2002. Theory, development, and applicability of the surface water hydrologic model CASC2D. Hydrol. Process. 16, 255–275.
Kouwen, N., 2018. Canadian Hydrological And Routing Model. User manual. ENVIRONMENT CANADA (for surface routing see p. 2-14). Also see: http://www.civil.uwaterloo.ca/watflood/index.htm.
FLO-2D, 2003. FLO-2D User Manual. Nutrioso.
Bergström, S., 1992. The HBV model—its structure and applications. SMHI RH No 4. Norrköping.
Seibert, J., 2005. HBV Light Version 2 User's Manual. Department of Physical Geography and Quaternary Geology, Stockholm University.
Feldman, A.D. (Ed.), 2000. Hydrologic Modelling System HEC-HMS Technical Reference Manual. US Army Corps of Engineers, Hydrologic Engineering Centre, Washington, DC.
Burek, P., Van der Knijff, J., De Roo, A., 2013. LISFLOOD distributed water balance and flood simulation model, revised user manual. JRC Technical Report 'EUR 26162 EN'.
DHI, 2017b. MIKE 11 A modelling system for rivers and channels. Reference Manual, DHI.
Van Beek, L.P.H., Bierkens, M.F.P., 2008. The Global Hydrological Model PCR-GLOBWB: Conceptualization, Parameterization and Verification, Report. Department of Physical Geography, Utrecht University, Utrecht, The Netherlands. http://vanbeek.geo.uu.nl/suppinfo/vanbeekbierkens2009.pdf.

Markstrom, S.L., Regan, R.S., Hay, L.E., Viger, R.J., Webb, R.M.T., Payn, R.A., LaFontaine, J.H., 2015. PRMS-IV, the Precipitation-Runoff Modeling System, Version 4, Techniques and Methods 6–B7. U.S. Geological Survey, Reston, Virginia.

Markstrom, S.L., Niswonger, R.G., Regan, R.S., Prudic, D.E., Barlow, P.M., 2008. GSFLOW—Coupled Ground-water and Surface-water FLOW model based on the integration of the Precipitation-Runoff Modeling System (PRMS) and the Modular Ground-Water Flow Model (MODFLOW-2005): U.S. Geological Survey Techniques and Methods 6-D1. 240 p.

Sayama, T., 2017. Rainfall-Runoff Inundation (RRI) Model, Version 1.4.2. International Centre for Water Hazard and Risk Management (ICHARM) and Public Works Research Institute (PWRI), Japan.

Quick, M.C., Pipes, A., 1977. U.B.C. watershed model. Hydrol. Sci. J. 22 (1), 153–161.

Quick, M.C., 1995. The UBC watershed model. In: Singh, V.P. (Ed.), Computer Models of Watershed Hydrology. Water Resources Publications, Colorado.

Brooks, R.H., Corey, A.H., 1964. Hydraulic properties of porous media. Hydrology Papers, No. 3. Colorado State University.

Liang, X., Lettenmaier, D.P., Wood, E.F., Burges, S.J., 1994. A simple hydrologically based model of land surface water and energy fluxes for general circulation models. J. Geophys. Res. 99 (D7), 14415–14428.

Gao, H., Tang, Q., Shi, X., Zhu, C., Bohn, T.J., Su, F., Sheffield, J., Pan, M., Lettenmaier, D.P., Wood, E.F., 2010. Water budget record from variable infiltration capacity (VIC) model. In: Algorithm Theoretical Basis Document for Terrestrial Water Cycle Data Records. UNSPECIFIED. Available for download from https://eprints.lancs.ac.uk/id/eprint/89407.

Lohmann, D., Raschke, E., Nijssen, B., Lettenmaier, D.P., 1998. Regional scale hydrology: I. Formulation of the VIC-2L model coupled to a routing model. Hydrol. Sci. J. 43 (1), 131–141.

Schulla, J., 2012. Model description WaSiM. Hydrology Software Consulting J. Schulla, Zurich.

Zhao, R.-J., 1992. The Xinanjiang model applied in China. J. Hydrol. 135, 371–381.

Yao, C., Li, Z., Yu, Z., Zhang, K., 2012. A priori parameter estimates for a distributed, grid-based Xinanjiang model using geographically based information. J. Hydrol. 468–469 (2012), 47–62. https://doi.org/10.1016/j.jhydrol.2012.08.025.

Fang, Y.-H., Zhang, X., Corbari, C., Mancini, M., Niu, G.-Y., Zeng, W., 2017. Improving the Xin'anjiang hydrological model based on mass–energy balance. Hydrol. Earth Syst. Sci. 21, 3359–3375. https://doi.org/10.5194/hess-21-3359-2017.

CHAPTER

7 Models of river flow

7.1 What does a river flow model do?

We discussed earlier that rivers are an integral part of a catchment, and so to model water flow processes in a catchment we must also model the rivers in the catchment, one way or another. However, you may have noticed that some catchment hydrological models may not have an explicit river model component. So, an obvious question is when do we need a river model within a catchment model and when we do not? Some of these questions are discussed in Chapter 1. A more relevant question for this chapter is what is the role of a river model in a catchment and how does a catchment model handle river flow?

We know (as discussed in Chapters 4–6) that rainwater falling on a catchment may reach the rivers by three ways: through surface flow (also called overland flow or direct flow), through interflow (that is the flow directly from the unsaturated zone, also called lateral flow or subsurface stormflow) and flow from the saturated zone or groundwater (also commonly referred to as baseflow). The purpose of a river model is to route all the water that enters the river channel to further downstream all the way to the outlet of the entire catchment represented in the model. The key phrase here is "to route water through the channel", and so river flow modelling is also commonly referred to as river flow routing. In this sense, it is useful to understand the flow routing with respect to a hydrograph (discharge, Q, verses time, T). The essence of a flow routing is to estimate a hydrograph at a section downstream on a river from a known hydrograph at a section upstream and water that may enter into the river reach between the upstream and downstream sections, usually referred to as lateral flow. Note that this meaning of 'lateral flow' is not the same as that sometimes used as an alternative name for the interflow. When a hydrograph is routed through the river channel (upstream to downstream), the shape of the hydrograph changes. The changes in the hydrograph can be described by two measures: attenuation and translation (or lag). Attenuation refers to the change (decrease) in the peak magnitude of the hydrograph when routed from upstream to downstream. Translation is the time lag between the time of occurrence of the peak discharge at the upstream and downstream hydrographs. By plotting the two hydrographs together, we can observe these changes (Fig. 7.1).

Note that river models or river modelling software usually refer to models that are specifically for rivers only, which are separate from catchment or hydrological models. These models are also called river hydraulic models, and we may also call

Catchment Hydrological Modelling. https://doi.org/10.1016/B978-0-12-818337-3.00005-2

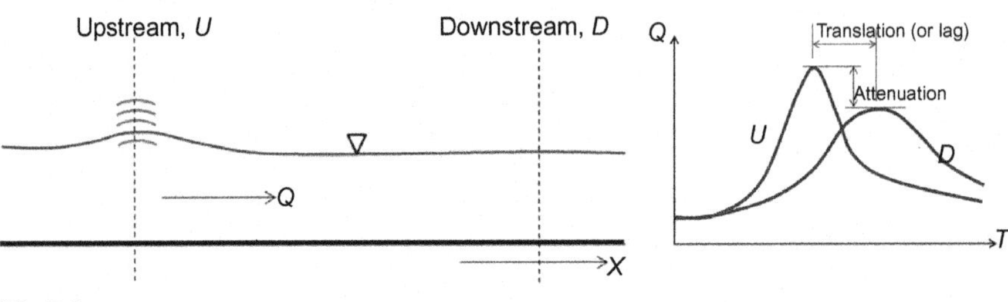

FIG. 7.1

Illustration of upstream and downstream hydrographs. On the left is a schematic longitudinal section of a river channel with upstream and downstream points indicated, and on the right are two hypothetical hydrographs: upstream (*U*) and downstream (*D*). The short curve lines at the upstream section (left graph) are shown to indicate water level/discharge has begun to rise, the effect of which is yet to be felt at the downstream section.

them "specialist" river models in the same way we may call "specialist" groundwater models (Chapter 6). When a river is modelled within a catchment model, river discharge computation is the main purpose, but in the "specialist" (hydraulic) river model, it is also about velocity and water depth and how river flow interacts with river bed, banks and river structures.

Catchment models can be classified as lumped, semidistributed or distributed in terms of representing spatial variation of catchment characteristics and processes (discussed in Chapter 1). In a fully lumped model, rivers are usually not represented as a distinct component, and in that case the sum of the runoff components (e.g., Q_{sf}, Q_{if}, and Q_{gw}) with their own transfer functions are treated as the total runoff from the catchment.

In a semidistributed model structure, in which the entire catchment is divided into a number of subcatchments, rivers are usually represented. Each subcatchment may have one river reach, and each river reach is connected to a downstream reach to become a part of the river network in the catchment. The total runoff from a subcatchment is drained into the river reach in the same subcatchment. If a subcatchment does not have its own river reach, it will drain to its immediate downstream river reach in the river network. The river model component then routes the runoff inputs from subcatchments through the river network to the outlet of the catchment. In some cases, a semidistributed model may be applied without a distinct representation of a river or river network. In such an application, transfer functions may be applied to the runoff from each subcatchment to estimate the runoff at the outlet of the entire catchment.

In a distributed model structure, the catchment area is divided into smaller areas using usually square or rectangular grids. A grid cell that has a river is indicated as a 'river cell' and each river cell is connected to a downstream river cell to form a river network. Runoff generated from grid cells in the catchment is transferred to the nearest river cell either based on a transfer function or based on some kind of flow routing. See Chapter 5 for surface flow routing and Chapter 6 for groundwater flow. The river flow component routes the flow through the river network formed by the river cells.

The connections of a river flow component to other components of a catchment model is presented in Fig. 7.2. As described above, how a river flow component is represented in different catchment models varies. The connections shown here is more comprehensive than that may be found in some models. In other words, not all models represent all the connections shown in the figure. The connections 2, 4 and 6 are the major connections, which represent the surface flow, interflow and groundwater flow inputs to the river routing component from surface flow, unsaturated zone and groundwater flow models, respectively. The connections 3, 5 and 7 are in the other direction (that is flow from river to these three components). These are shown with dotted lines to suggest that these connection representations may be found only in some, generally comprehensive, models. Connections 1 and 8 are also represented with dotted lines, because direct evaporation from river water (connection 1) and direct snowmelt input to river flow (connection 8) are simulated only in some models.

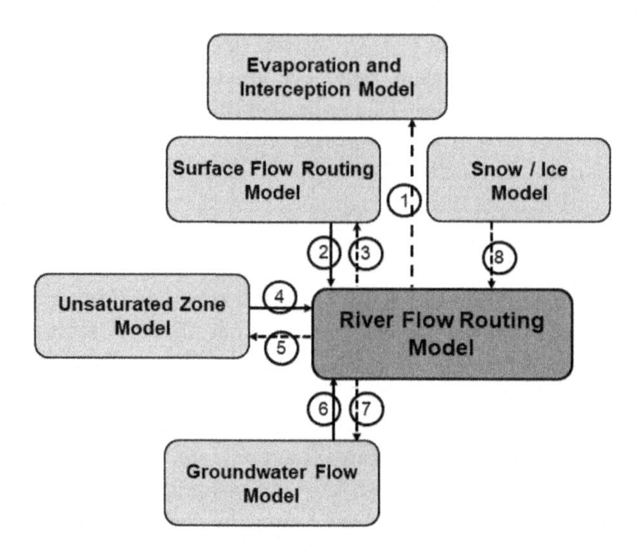

FIG. 7.2

Connections of the River Flow Routing component to other components of the catchment model. Connections indicated by the numbers represent: (1) Evaporation from river water (open water evaporation), (2) Surface runoff (Q_{sf}) input to the river, (3) Overbank flow to the floodplain, which may become part of surface flow, (4) Interflow (Q_{if}) input from the unsaturated zone, (5) Infiltration from river bed to the unsaturated zone in parts of the river reach not linked with the saturated zone, (6) Groundwater input to the river (Q_{gw}), (7) Recharge to the groundwater from the river, (8) Direct snowmelt input to the river, if exists. Note that not all hydrological models represent all the connections shown here.

7.2 Computation of steady-uniform flow in a river channel

Flow conditions in an open channel can be steady or unsteady and uniform or non-uniform. Flow condition is said to be steady when flow depth and velocity (and therefore discharge) remain constant over a period of time, that is $\partial h/\partial t = 0$, or $\partial V/\partial t = 0$. Conversely, the flow is in unsteady condition when depth, velocity or discharge is changing over a period of time, that is $\partial h/\partial t \neq 0$, or $\partial V/\partial t \neq 0$. The condition for uniform or non-uniform is whether the flow characteristics, e.g. depth and velocity, are same or varied with distance along the channel. That is, the flow condition is said to be uniform if $\partial h/\partial x = 0$, or $\partial V/\partial x = 0$, where x is the distance between the two sections. Conversely, in a non-uniform flow $\partial h/\partial x \neq 0$, or $\partial V/\partial x \neq 0$. In a given channel section with constant channel characteristics, if the discharge through it remains constant, the flow condition remains steady.

If we consider a small length of a river, and observe water depths at two sections (separated by a small distance) along the direction of flow, and suppose that we find the water depths are different, but both depths are constant over time. This is the situation we call the flow condition steady but non-uniform, so it is steady non-uniform flow. If the depths at both sections are also same, and also constant over time then it is uniform and steady, or in short steady-uniform flow.

In the hydrographs shown in Fig. 7.1, we see that initially the discharge in each hydrograph is more or less constant. After a certain time, the discharge starts rising, reaches a peak value and recedes. After a certain time it may retain a constant discharge again. So, there is clearly a period observed in the hydrographs when the flow condition is unsteady. Typically, such a rise in discharge is caused by a sudden increase in the runoff input due to a rainfall event. A non-uniform flow condition occurs where the channel characteristics, such as bed slope, width, or roughness, change from one section to another.

In a river channel with a constant discharge, the steady-uniform condition can be found in a channel reach with constant bed slope, uniform cross-section and uniform bed roughness. When the flow condition is steady as well as uniform (that is $\partial h/\partial t = 0$ and $\partial h/\partial x = 0$), hydraulic computations are a lot easier. As a result, velocity and discharge computation formulas derived for a steady-uniform condition are widely used. Two popular steady-uniform flow equations are Chezy equation (Eq. 7.1) and Manning equation (Eq. 7.2). Both Chezy and Manning equations relate the average velocity (V) through a channel to three parameters, namely the slope of the channel bed (S_0), channel hydraulic radius (R) and an empirical factor to represent the bed resistance to the flow, also called bed roughness.

$$V = C\sqrt{RS_0} \tag{7.1}$$

$$V = \frac{1}{n}R^{\frac{2}{3}}\sqrt{S_0} \tag{7.2}$$

where V is in m/s and R is in m. The bed slope $S_0 = \Delta z/\Delta x$, where Δz (m) is the difference in bed elevation between the two sections and Δx (m) is the horizontal distance between the sections (Fig. 7.3). Note here that for small slopes (mostly

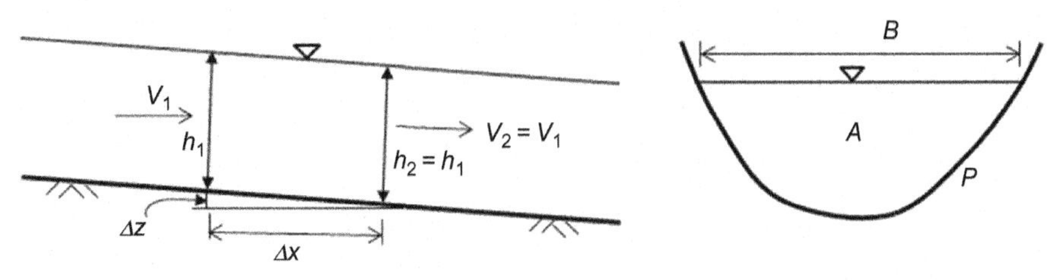

FIG. 7.3

A channel longitudinal section (left) and a lateral cross-section (right) in a steady-uniform condition (with $h_1 = h_2$ and $V_1 = V_2$).

encountered in river channels), the distance along the slope (say, Δs) is assumed to be equal to the horizontal distance, that is $\Delta s \approx \Delta x$. As we can see, while the resistance factor C (Chezy equation) is directly proportional to the flow velocity V, the roughness coefficient n (Manning equation) is inversly related. A larger n value represents a rougher bed condition and higher resistance to flow, while a larger C value represents a smoother bed condition and lower resistance to flow. To balance the units in the formulas (Eqs 7.1 and 7.2), C and n are also specified with units $m^{1/2}\,s^{-1}$, and $m^{-1/3}\,s$, respectively.

The hydraulic radius R is defined by area (A) of the channel cross-section up to the water depth divided by its perimeter (P), that is $R = A/P$. To distinguish the A and P up to the water level from those of the entire channel cross-section, they are commonly referred to as 'wetted area' and 'wetted perimeter'.

Substituting discharge $Q = AV$ in Eqs (7.1) and (7.2) the Chezy and Manning equations can be written for discharge as

$$Q = CA\sqrt{RS_0} \tag{7.3}$$

$$Q = \frac{1}{n}AR^{\frac{2}{3}}\sqrt{S_0} \tag{7.4}$$

where Q is in m^3/s and A in m^2. In the case of a composite channel consisting of the main channel and floodplains, the uniform flow discharge through the channel are estimated applying the formula (Eq. 7.3 or 7.4) separately for the main channel and the floodplain. In river modelling, the empirical parameters n and C are usually considered as calibration parameters. Detailed tables with values of n for different channel bed conditions can be found in Chow (1959) and (Brunner, 2016).

7.3 Conservation of mass (continuity) equation for a river flow model

The conservation of mass (or volume in case of constant density) is arguably the most important concept in catchment modelling. The catchment water balance equation we discussed in Chapter 1 is also the result of conservation of water volume. A river channel has a defined geometry represented by width, depth, length and slope, and so the conservation of mass equation for a river channel can be represented in somewhat different form than we do for the catchment as a whole. For simplicity, consider a one-dimensional flow and assuming a small length Δx of the channel, the continuity equation tells that 'the difference in the discharge inflow into and outflow from the channel length Δx must be balanced by the change in the water stored in the channel within this length'. For an elemental length Δx of uniform cross-section, the change in storage can be replaced by the change in area occupied by the water in the channel. Which means in differential form we can write,

$$\frac{\partial Q}{\partial x} + \frac{\partial A}{\partial t} = 0 \tag{7.5}$$

For further explanation of this equation, suppose over a certain time we find that the total discharge inflow (Q_{in}) is more than the outflow (Q_{out}). Which means more water is 'in' than 'out' over that time period, and assuming no lateral flow, the total volume difference due to the difference in Q_{in} and Q_{out} ($Q_{in} > Q_{out}$) must be accommodated within the channel reach (length Δx) by increasing the wetted area (cross-section area occupied by the water). From the same logic, if $Q_{in} < Q_{out}$, the difference must be compensated by decreasing the wetted area. Note that in the above equation, the term $\partial Q/\partial x$ is positive when Q is increasing downstream and $\partial A/\partial t$ is positive when A is increasing over time.

If water is added to or removed from the channel within the length Δx, which is commonly referred to as lateral flow, that needs to be included in the continuity equation. The continuity equation with lateral flow is used in Chapter 5 with the kinematic wave method. For convenience, it is recalled here (Eq. 7.6):

$$\frac{\partial Q}{\partial x} + \frac{\partial A}{\partial t} = q_l \tag{7.6}$$

where q_l (m^2/s) is lateral inflow into the channel reach and taken positive. If water is removed from the channel reach (lateral outflow), then q_l will be negative. For example, in Fig. 7.2, the connections 1, 5 and 7, if included, are typically lateral outflow (thus $-q_l$). Eqs (7.5) and (7.6) are called the conservative form of the continuity equation, with A being the conserved variable (Guinot, 2008). The nonconservative form is also shown in Chapter 5 with the kinematic wave method.

7.4 River flow routing methods

Flow routing methods refer to mathematical techniques to simulate the water movement in a river channel. The routing methods are also commonly referred to as flood wave propagation. The primary governing principles of flow routing methods are the conservation of mass and conservation of momentum. The conservation of momentum concerns the balance of forces. In water flow in a river channel, forces considered are the inertia force (defined by the Newton's second law of motion), gravity force, pressure force and shear force due to bed resistance to the flow. These forces are considered in the derivation of the Saint-Venant equations for dynamic routing of river flow. However, due to the complexity of implementing fully dynamic flow equations (consisting continuity and full momentum equations), different simplifications are used in catchment models depending on the required details and availability of data. Accordingly, a number of different flow routing methods are in practice. Here two groups of flow routing methods are presented. In the first group are the dynamic wave method (also called the Saint Venant equations) and two approximations (diffusive wave and kinematic wave) of the full dynamic wave method. All three are widely used methods for river flow modelling. In the second group are the Muskingum method and its extension Muskingum-Cunge method. Both are popular methods particularly with catchment hydrological models.

7.4.1 Dynamic wave method: the Saint Venant equations

The Saint Venant equations are dynamic wave equations to simulate unsteady flow in river channels in one-dimension. Two equations represent the Saint Venant equations, one for the conservation of mass and one for the conservation of momentum. The conservation of mass (or continuity) equation is represented by Eq. (7.5) or (7.6). To recall, the conservation of momentum is the balance of forces. To understand the force terms that are considered in the derivation of the Saint Venant momentum equation, let us consider a simplified open channel flow profile shown in Fig. 7.4. The forces acting on the control volume of elementary length Δx are as follows.

Momentum force (F_m) consisting of local acceleration ($\partial V / \partial t$) and convective acceleration ($\partial V / \partial x$):

$$F_m = \rho\left(\frac{\partial V}{\partial t} + V\frac{\partial V}{\partial x}\right)\Delta x A \tag{7.7}$$

where ρ is the density of water (kg/m^3).

Pressure force (F_p):

$$F_p = -\rho g \frac{\partial h}{\partial x}\Delta x A \tag{7.8}$$

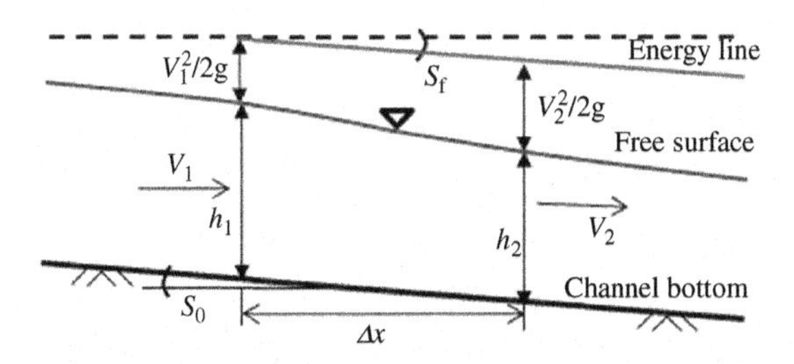

FIG. 7.4

A simplified free surface (open channel) flow profile with notations for the derivation of the Saint Venant momentum equation.

where g is gravity (m/s^2). The negative sign indicates that a positive gradient $\partial h/\partial x$ (i.e., $h_2 > h_1$) results in the net pressure force in the negative x-axis or negative flow direction.

Shear (friction) force (F_s) is defined with respect to the slope of the energy line, or energy slope (S_f) (Fig. 7.4):

$$F_s = -\rho g \Delta x A S_f \tag{7.9}$$

The negative sign indicates that the direction of friction force is opposite to the flow.

Gravity force (F_g):

$$F_g = \rho g \Delta x A S_0 \tag{7.10}$$

Note that same as in the Manning and Chezy equations, the bed slope $S_0 = \Delta z/\Delta x$ is used assuming a small bed slope angle, such that $\sin\theta \approx \tan\theta$, where θ is the angel of the channel bed slope.

The energy slope is commonly represented by the Chezy and Manning equations given by

$$S_f = \frac{V^2}{C^2 R} \tag{7.11}$$

$$S_f = \frac{n^2 V^2}{R^{4/3}} \tag{7.12}$$

Equating the momentum force (Eq. 7.7) to the other three forces (pressure force, shear force and gravity force, that is, Eqs 7.8–7.10), we obtain

$$\rho\left(\frac{\partial V}{\partial t} + V\frac{\partial V}{\partial x}\right)\Delta x A = -\rho g \frac{\partial h}{\partial x}\Delta x A - \rho g A \Delta x S_f + \rho g \Delta x A S_0 \tag{7.13}$$

Dividing both sides by $\rho \Delta x A$ and rearranging, we obtain the Saint Venant momentum equation for one-dimensional (1D) free surface flow

$$\frac{\partial V}{\partial t} + V\frac{\partial V}{\partial x} + g\frac{\partial h}{\partial x} - g(S_0 - S_f) = 0 \tag{7.14}$$

This equation is also commonly written using Q (with $V = Q/A$) as

$$\frac{\partial Q}{\partial t} + \frac{\partial}{\partial x}\left(\frac{Q^2}{A}\right) + gA\frac{\partial h}{\partial x} - gA(S_0 - S_f) = 0 \tag{7.15}$$

The five terms of the momentum Eq. (7.14) or (7.15) from left to right represent forces due to local acceleration, convective acceleration, pressure, gravity and friction, respectively. More detailed steps and explanation involved in the derivation of Saint Venant equations can be found in many text books on open channel hydraulics or related topics, e.g., Chow et al. (1988), Chanson (2004) and Guinot (2008).

The application of the Saint Venant equations (for fully dynamic flow routing) involves solving Eq. (7.5) or (7.6) and Eq. (7.15). These are relatively complicated partial differential equations and different numerical techniques are used to solve them. Some examples of numerical techniques used for solving the Saint Venant equations can be found in Chow et al. (1988) and in the reference manual of HEC-RAS software (Brunner, 2016). These equations are commonly used in 'specialist' river modelling tools., e.g., HEC-RAS (Brunner, 2016) and MIKE 11 (DHI, 2017a). However, for river flow routing within a catchment (hydrological) model, the full Saint Venant equations are rarely used. Instead, in a catchment model, approximations of these equations (Section 7.4.2) are commonly used.

7.4.2 Diffusive wave and kinematic wave approximations

With the Saint Venant momentum equation presented above (Eq. 7.14 or 7.15), two approximations are commonly used in practice: neglecting the acceleration terms, that is $\partial V/\partial t$ and $V(\partial V/\partial x)$, and neglecting both the acceleration terms and pressure force term, $g(\partial h/\partial x)$. The former is called the 'diffusive wave' approximation and the latter 'kinematic wave' approximation.

Thus, the diffusive wave routing method consists of the continuity Eq. (7.5) or (7.6), and momentum equation of the form

$$\frac{\partial h}{\partial x} - (S_0 - S_f) = 0 \tag{7.16}$$

Similarly, the kinematic wave routing method consists of the continuity Eq. (7.5) or (7.6) and the momentum equation

$$S_0 - S_f = 0 \tag{7.17}$$

Eq. (7.17) means that the channel bed slope is assumed to be equal to the energy slope, which is also the condition assumed in the Chezy and Manning formulas. That means that, in the kinematic wave method, the continuity equation combined with the uniform flow formula (e.g., Manning or Chezy) is used.

In Chapter 5, both diffusive wave and kinematic wave methods and their solution techniques are discussed in connection with the surface flow routing. The same numerical techniques and solution procedures can be applied for the river flow routing. In the case of diffusive wave routing, the equations described in Chapter 5 are for a two-dimensional (2D) formulation, which is often preferred for surface flow routing. For river flow routing, a 1D formulation is more commonly used.

7.4.3 Muskingum method

The Muskingum method is one of the earliest and widely used methods for river flow routing with conceptual hydrological models. The basic idea of a Muskingum method is to estimate a hydrograph at a section downstream from a known hydrograph somewhere upstream of the river without explicit use of the river length and geometry in the computation. The starting point for the Muskingum method is also the conservation of mass or volume. Which means that between two sections (upstream and downstream) of a river reach (as shown in Fig. 7.1), the change in storage volume S over time t must be balanced by the difference between the inflow discharge from upstream (Q_U) and outflow discharge at downstream (Q_D). That is

$$\frac{dS}{dt} = Q_U - Q_D \tag{7.18}$$

Eq. (7.18) can be rewritten with a simple finite difference approximation as

$$\frac{S^{n+1} - S^n}{\Delta t} = \frac{Q_U^n + Q_U^{n+1}}{2} - \frac{Q_D^n + Q_D^{n+1}}{2} \tag{7.19}$$

where n and $n+1$ represent time levels current and one-time step later, respectively. To solve this equation one-time step at a time, the upstream discharges are known for both n and $n+1$ time levels, and the storage and downstream discharge are known for n time level. There remain two unknowns: the storage and downstream discharge for time level $n+1$, so we need a second equation. For this, the Muskingum method defines a relationship between storage and discharge using two coefficients, K and X, in the form

$$S = KQ_D + KX(Q_U - Q_D) \tag{7.20}$$

The coefficients K and X are empirical parameters. K has the dimension of time and represents the travel time of the flood wave through the channel reach (Chow et al., 1988). X is a dimensionless weighting factor with its value between 0 and 0.5. The storage terms KQ_D and KX $(Q_U - Q_D)$ are also referred to as prism and wedge storages, respectively.

Substituting S from Eq. (7.20) to (7.19) and rearranging, we obtain

$$Q_D^{n+1} = C_1 Q_U^n + C_2 Q_U^{n+1} + C_3 Q_D^n \tag{7.21}$$

The coefficients C_1, C_2 and C_3 are related to parameters K and X and the time step of computation (Δt), given by

$$\left.\begin{aligned} C_1 &= \frac{\Delta t + 2KX}{2K(1-X) + \Delta t} \\ C_2 &= \frac{\Delta t - 2KX}{2K(1-X) + \Delta t} \\ C_3 &= \frac{2K(1-X) - \Delta t}{2K(1-X) + \Delta t} \end{aligned}\right\} \tag{7.22}$$

Note that $C_1 + C_2 + C_3 = 1$. Thus, we can see in Eq. (7.21) that the outflow (downstream) discharge at time level $n+1$ is the weighted sum of the inflow (upstream) discharges at time levels n and $n+1$ and the downstream discharge at time level n.

As said earlier, K and X are empirical parameters, which means that we need observation data at both ends (input and output) to estimate their values. In other words we need inflow and outflow hydrographs to estimate these parameters. When used in a catchment model, their values are usually adjusted by calibration. Because these parameters are dependent on hydrographs at upstream and downstream, in principle their values should be calibrated for each subcatchment.

7.4.4 Muskingum-Cunge method

The Muskingum-Cunge method is a modification of the Muskingum method proposed by Cunge (1969) and further explained by Ponce and Yevjevich (1978) and Ponce (1986). It improves the Muskingum method in two ways. First, it defines the expressions for parameters K and X (which are empirical in the Muskingum method) using physically meaningful parameters. Second, the Muskingum method does not have space dimension, that is the length between the inflow and outflow points are not included. The Muskingum-Cunge method includes both time and space dimensions, similar to the kinematic wave method, and also uses the channel geometry through the Manning or Chezy equation. In this sense the Muskingum method is a lumped method and Muskingum-Cunge is a distributed method. More about the advantages of the Muskingum-Cunge method compared to the Muskingum and kinematic wave methods can be found in Chow et al. (1988) and NRCS (2014).

Using the space and time discretization shown in Fig. 7.5, the Muskingum-Cunge equation for discharge can be written as (see, e.g., Feldman, 2000; Ponce, 1986)

$$Q_j^{n+1} = C_1 Q_{j-1}^n + C_2 Q_{j-1}^{n+1} + C_3 Q_j^n + C_4 (q_l \Delta t) \tag{7.23}$$

where q_l (m^2/s) is the lateral inflow (same as that defined in Eq. 7.6). The coefficients C_1, C_2 and C_3 are the same as in Muskingum method defined by Eq. (7.22), but with

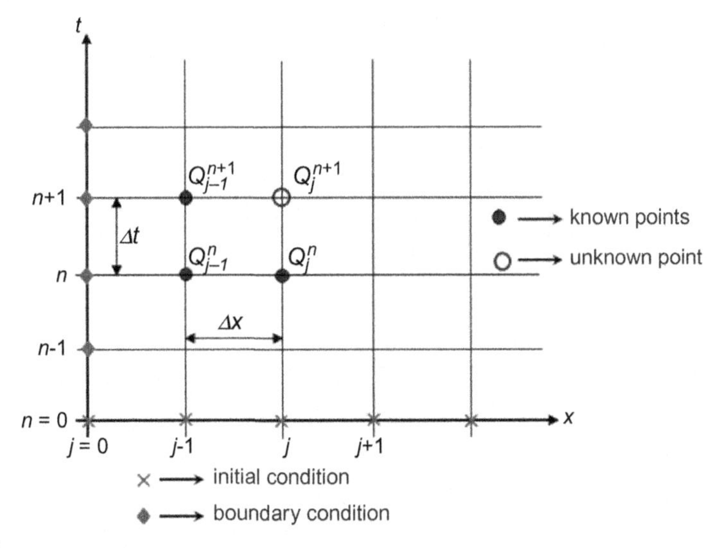

FIG. 7.5

Notations used for the space–time discretization for the Muskingum-Cunge routing method given by Eq. (7.23). The notations are the same as used for the kinematic wave routing presented in Chapter 5, Fig. 5.4.

differently estimated K and X values. The coefficient C_4 is added when the lateral flow is considered (Ponce, 1986), and its value is given by

$$C_4 = \frac{2\Delta t}{2K(1-X)+\Delta t} \tag{7.24}$$

The parameters K and X are given by

$$\left.\begin{aligned} K &= \frac{\Delta x}{c} \\ X &= \frac{1}{2}\left(1 - \frac{Q}{BS_0 c\Delta x}\right) \end{aligned}\right\} \tag{7.25}$$

where c is the wave celerity (m/s), B is the width (m) of the water surface, S_0 is the channel bed slope and Δx is the channel segment length used for the computational step (m). The wave celerity is defined by

$$c = \frac{\partial Q}{\partial A} \tag{7.26}$$

The relationship between Q and A can be found using the Chezy or Manning equation same as in the case of kinematic wave routing described in Chapter 5. See also Ponce (1986) and NRCS (2014). At any time step, the value A (and in case of a non-rectangular channel also the top width) is dependent on the discharge, which is to be computed (the discharge at section j at time level $n+1$). Therefore, a reference estimate value is used, typically as an average of the three known discharge values, that are $Q\,(j-1, n)$, $Q\,(j-1, n+1)$ and $Q\,(j, n)$. More options for the reference Q are discussed in Ponce and Yevjevich (1978).

Note that the Muskingum-Cunge method given by Eq. (7.23) is not an implicit method and so care needs to be taken in selecting time and space steps to ensure stability of the numerical solution (see, e.g., Feldman, 2000).

7.5 How different catchment models treat river flow?

As we have seen there are various approaches and methods to model the river flow component. Methods and approaches used in 17 catchment models for river flow routing are reviewed and briefly described here (Table 7.1). The selected models are among the widely used models, but the list is not exhaustive, and the purpose of presenting the table is not for giving author's judgement about the model. It is simply intended to describe the key concepts of the methods/approaches used in the models. The descriptions attempt to capture the main essence or feature, but may not be detailed enough to provide complete details of the methods, for which the referenced literature should be consulted.

Table 7.1 River flow routing methods in various hydrological models.

Modelling software	Method available	Additional information
BROOK90	It does not have river flow routing. It is a lumped model without horizontal transfer of water (Federer, 2002; Federer et al., 2003).	
CASC2D	It has two options available: one with the diffusive wave equation solved with an explicit numerical scheme, and one with the fully dynamic equation solved with an implicit numerical scheme. Different types of channel cross-sections are allowed (Ogden, 2001).	
CHARM (also WATFLOOD)	Flow routing in a river channel is based on a simple "storage routing" method, which uses simple volume balance in the river reach, combined with the Manning's uniform flow formula. It assumes a rectangular channel cross-section for the main channel and triangular for the side channels (left/right) for overbank flow (Kouwen, 2018).	CHARM: Canadian Hydrological and Routing Model
Flo2D	It uses the 1D dynamic wave equation for the river channel flow and allows cross sections defined by up to 8-points of variable area or rectangular or trapezoidal (FLO-2D, 2003).	
HBV (and HBV-light)	It does not have a river routing as such, but uses a transform function to estimate the catchment runoff from the runoff generated from the soil and groundwater layers (Bergström, 1992; Nielsen and Hansen, 1973; Seibert, 2005). To apply the model as (semi) distributed (i.e., dividing the entire catchment into subcatchment areas), routing between subcatchments is necessary and simple methods, e.g., Muskingum routing, may be applied (Lindström et al., 1997).	See Lindström et al. (1997) for some modifications applied for a semidistributed version (HBV-96).

Continued

Table 7.1 River flow routing methods in various hydrological models—cont'd

Modelling software	Method available	Additional information
HEC-HMS	Several methods are available varying from simple lag method and Muskingum routing to kinematic wave and Muskingum-Cunge routing (Feldman, 2000). Note that HEC-HMS and HEC-RAS (River Analysis System) (Brunner, 2016) can also be integrated, which is particularly useful for river reaches where overbank flow and inundation are expected. HEC-RAS is a river hydraulic model and allows fully dynamic wave routing for river flow.	
LISFLOOD	It uses the kinematic wave method implemented with an implicit numerical scheme (Burek et al., 2013).	
MIKE SHE	It integrates the MIKE Hydro River (or MIKE 11) for the river flow, which allows 1D fully dynamic routing (DHI, 2017b).	
NAM	When applied as a lumped rainfall-runoff model it does not use river routing. But NAM is also integrated with MIKE 11, where it can be used as semidistributed model (with subcatchments) and the river flow is modelled with MIKE 11. MIKE 11 has options for kinematic, diffusive as well as fully dynamic routing methods (DHI, 2017a).	
PCR-GLOBWB	It uses the kinematic wave method with a rectangular channel. An explicit numerical scheme is used but with a variable time step to keep the numerical solution stable (Van Beek and Bierkens, 2009).	
PRMS also used in GSFLOW	It has the option to use Muskingum routing with defined stream segments, but a simpler option, e.g., stream flow = sum of the surface, interflow and groundwater runoff components, is also allowed (Markstrom et al., 2015).	PRMS: Precipitation Runoff Modelling System.

Table 7.1 River flow routing methods in various hydrological models—cont'd

Modelling software	Method available	Additional information
RRI model	It uses the 1D diffusive wave method with a rectangular channel (Sayama, 2017).	
SWAT	Two methods are available: the "variable storage routing" method and Muskingum method, with a trapezoidal cross-section for the main channel and Manning's equation for average velocity in the channel (Neitsch et al., 2011).	
UBC model	It has no river flow routing component in the model as such, but lake or reservoir routing from a catchment is available (Quick and Pipes, 1977).	UBC: University of British Colombia.
VIC	In the more recent versions of VIC (e.g., described in Gao et al., 2010), streamflow routing is based on "linearized Saint-Venant equations", as in RVIC (Hamman et al., 2017) and included in VIC version 5 (Hamman et al., 2018).	VIC (Variable Infiltration Capacity). See also the VIC version 5 webpage: https://vic.readthedocs.io/en/master/
WaSiM	The method used for the channel flow routing may be loosely described as a sort of storage routing "with different velocities for different water levels" which are based on the Manning-Strickler equation. It describes the method as a three-step process: estimate translation for the channels, apply the storage approach, and superimpose discharges from different subbasins (Schulla, 2021).	
Xinanjiang model	In the original version (Zhao, 1992), river reaches are not specifically represented but the discharges from subcatchments to the catchment outlet are routed using the Muskingum method. In the grid-based version (Yao et al., 2012), routing in the channel is based on the kinematic wave equations.	

References

Bergström, S., 1992. The HBV model—its structure and applications. SMHI RH No 4. Norrköping.

Brunner, G.W., 2016. HEC-RAS River Analysis System, Hydraulic Reference Manual, Version 5.0. US Army Corps of Engineers, Hydrologic Engineering Centre, Davis, CA.

Burek, P., Van der Knijff, J., De Roo, A., 2013. LISFLOOD distributed water balance and flood simulation model, revised user manual. JRC Technical Report 'EUR 26162 EN'.

Chanson, H., 2004. The Hydraulics of Open Channel Flow: An Introduction, second ed. Elsevier.

Chow, V.T., 1959. Open-Channel Hydraulics. McGraw-Hill, Inc., Singapore. International Edition 1973.

Chow, V.T., Maidment, D.R., Mays, L.W., 1988. Applied Hydrology, International Edition. McGraw-Hill, Singapore.

Cunge, J.A., 1969. On the subject of a flood propagation method (Muskingum method). J. Hydraul. Res. 7, 205–230.

DHI (2017a). MIKE 11 A Modelling System for Rivers and Channels, Reference Manual. DHI, Denmark. https://manuals.mikepoweredbydhi.help/2017/Water_Resources/Mike_11_ref.pdf; Accessed on 31 July, 2021.

DHI (2017b). MIKE SHE Volume 2: Reference Guide. DHI, Denmark. https://manuals.mikepoweredbydhi.help/2017/Water_Resources/MIKE_SHE_Printed_V2.pdf; Accessed on 31 July, 2021.

Federer, C.A., 2002. BROOK 90: a simulation model for evaporation, soil water, and streamflow. http://www.ecoshift.net/brook/brook90.htm.

Federer, et al., 2003. Sensitivity of annual evaporation to soil and root properties in two models of contrasting complexity. J. Hydromereorol. 4, 1276–1290.

Feldman, A.D. (Ed.), 2000. Hydrologic Modelling System HEC-HMS. US Army Corps of Engineers, Hydrologic Engineering Centre, Davis, CA. Technical Reference Manual.

FLO-2D, 2003. FLO-2D User Manual. Nutrioso.

Gao, H., Tang, Q., Shi, X., Zhu, C., Bohn, T.J., Su, F., Sheffield, J., Pan, M., Lettenmaier, D.P., Wood, E.F., 2010. Water budget record from variable infiltration capacity (VIC) model. In: Algorithm Theoretical Basis Document for Terrestrial Water Cycle Data Records. UNSPECIFIED. Available for download from https://eprints.lancs.ac.uk/id/eprint/89407.

Guinot, V., 2008. Wave Propagation in Fluids: Models and Numerical Techniques. ISTE Ltd., London.

Hamman, J., Nijssen, B., Roberts, A., Craig, A., Maslowski, W., Osinski, R., 2017. The coastal streamflow flux inthe regional Arctic system model. J. Geophys. Res. Oceans 122, 1683–1701. https://doi.org/10.1002/2016JC012323.

Hamman, J.J., Nijssen, B., Bohn, T.J., Gergel, D.R., Mao, Y., 2018. The variable infiltration capacity model version 5 (VIC-5): infrastructure improvements for new applications and reproducibility. Geosci. Model Dev. 11, 3481–3496. https://doi.org/10.5194/gmd-11-3481-2018.

Kouwen, N., 2018. Canadian Hydrological and Routing Model. User manual. ENVIRONMENT CANADA. See also: http://www.civil.uwaterloo.ca/watflood/index.htm.

Lindström, G., Johansson, B., Persson, M., Gardelin, M., Bergström, S., 1997. Development and test of the distributed HBV-96 hydrological model. J. Hydrol. 201, 272–288. https://doi.org/10.1016/S0022-1694(97)00041-3.

Markstrom, S.L., Regan, R.S., Hay, L.E., Viger, R.J., Webb, R.M.T., Payn, R.A., LaFontaine, J.H., 2015. PRMS-IV, the Precipitation-Runoff Modeling System, Version 4, Techniques and Methods 6–B7. U.S. Geological Survey, Reston, VA.

Neitsch, S.L., Arnold, J.G., Kiniry, J.R., Williams, J.R., 2011. Soil and Water Assessment Tool Theoretical Documentation Version 2009. Texas Water Resources Institute. Available electronically from https://hdl.handle.net/1969.1/128050.

Nielsen, S.A., Hansen, E., 1973. Numerical simulation of the rainfall runoff process on a daily basis. Nordic Hydrol. 4 (3), 171–190. https://doi.org/10.2166/nh.1973.0013.

NRCS, 2014. National Engineering Handbook. Part 630 Hydrology, Chapter 17 Flood Routing. Natural Resources Conservation Service, US Department of Agriculture.

Ogden, F.L., 2001. A Brief Description of the Hydrologic Model CASC2D. Univ. Connecticut.

Ponce, V.M., 1986. Diffusion wave modeling of catchment dynamics. J. Hydr. Engrg. 112 (8), 716–727.

Ponce, V.M., Yevjevich, V., 1978. Muskingum-Cunge method with variable parameters. J. Hydraul. Div., ASCE 104 (HY12), 1663–1667.

Quick, M.C., Pipes, A., 1977. U.B.C. watershed model. Hydrol. Sci. J. 22 (1), 153–161.

Sayama, T., 2017. Rainfall-Runoff Inundation (RRI) Model, Version 1.4.2. International Centre for Water Hazard and Risk Management (ICHARM) and Public Works Research Institute (PWRI), Japan.

Schulla, J., 2021. Model description WaSiM, Version 10.06.00. Hydrology Software Consulting J. Schulla, Zurich. http://www.wasim.ch/downloads/doku/wasim/wasim_2021_en.pdf.

Seibert, J., 2005. HBV Light Version 2 User's Manual. Department of Physical Geography and Quaternary Geology, Stockholm University.

Van Beek, L.P.H., Bierkens, M.F.P., 2009. The Global Hydrological Model PCR-GLOBWB: Conceptualization, Parameterization and Verification. Report, Department of Physical Geography, Utrecht University, Utrecht, The Netherlands http://vanbeek.geo.uu.nl/suppinfo/vanbeekbierkens2009.pdf.

Yao, C., Li, Z., Yu, Z., Zhang, K., 2012. A priori parameter estimates for a distributed, grid-based Xinanjiang model using geographically based information. J. Hydrol. 468–469 (2012), 47–62. https://doi.org/10.1016/j.jhydrol.2012.08.025.

Zhao, R.-J., 1992. The Xinanjiang model applied in China. J. Hydrol. 135, 371–381.

CHAPTER

Models of snowmelt runoff 8

8.1 Role of snowmelt in a hydrological model

Many regions at high latitudes and high elevations receive precipitation as snowfall. When precipitation occurs as snow, it follows different dynamics of precipitation to runoff processes than rainfall. Unlike rain precipitation, snow precipitation remains as storage (accumulation) until the snowpack receives sufficient energy for melting and sublimation (latent heat). This delayed response of snowfall for precipitation to runoff generation results in a different hydrological regime in snow dominated catchments. Snowfall usually occurs as winter precipitation, which may remain as snow storage from several days to months. These winter snowfalls may be the major source of river water during spring and summer in those regions. Many river catchments in the Himalayan high-altitude regions, for example, depend on snow and glacier melt discharge for a significant part of the year. In some cold regions, some snow storage of a year may remain after the melt season (positive snow mass balance) and over time turn into ice mass. On the other hand, in other years, snow storage may melt faster and further deplete any ice storage from previous years (negative mass balance). With increasing global temperature, faster depletion of snow cover and glaciers has been an eminent threat.

In general, modelling a catchment with snow is more challenging than only rainfed catchments, not least because of more data requirements for modelling snowmelt runoff processes, lack of measurement data, and in mountainous catchments, in particularly, high variability of topography and weather. To simulate the snowmelt runoff in a hydrological model, broadly three steps can be distinguished. These steps are required to (1) determine the form of precipitation (rain or snow) on a particular time or day, (2) compute the snowmelt (determining when the melting starts and at what rate) through the energy exchange process in the snowpack, and (3) estimate the meltwater output from the snowpack, which involves snowpack mass balance. Note that steps 2 and 3 are intrinsically connected. These steps are described in Sections 8.2–8.4. Once the meltwater output is determined (step 3), the transport of the meltwater through surface and subsurface into the streams are dealt with in a catchment model in the same way as the transport of rainwater.

The snow (ice) melt model component is connected to three other components of the catchment model (Fig. 8.1): Evaporation and Interception, Unsaturated Zone and Surface Flow Routing. Evaporation including sublimation is part of the energy balance process (as latent heat transfer). Part of the canopy intercepted snow falls

Catchment Hydrological Modelling. https://doi.org/10.1016/B978-0-12-818337-3.00006-4

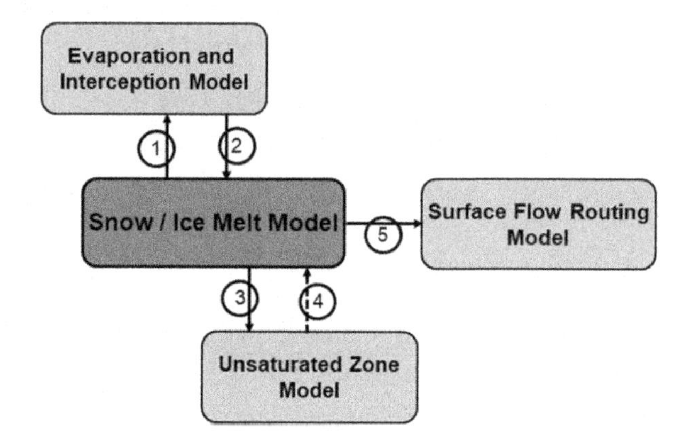

FIG. 8.1

Connections of the Snow (ice) melt model to other components of a catchment model. Connections indicated by the numbers represent: (1) Evaporation and sublimation estimates, (2) Throughfall of intercepted snow, (3) Infiltration input from the meltwater outflow and heat flux, (4) Ground heat flux from UZ soil to the snowpack, and (5) Direct runoff from the meltwater output for the surface flow routing.

on the snowpack (throughfall input). Meltwater output from the snowpack is divided into infiltration input to the unsaturated zone and direct runoff volume for the surface flow routing.

8.2 Determining the form of precipitation

Precipitation data are recorded as water equivalent and the information about the form of precipitation (snow or rain) is usually not available (Dingman, 2002). Unless such information is available, separation of precipitation into rain or snow for input to a catchment model is almost always based on the threshold temperature approach. For example, for precipitation data with daily time step, if the average temperate of a day is above a threshold temperature (which is usually set to 0°C), precipitation of that day is counted as rainfall. Although 0°C as the threshold temperature is commonly used, precipitation may fall as snow over a much wider range of temperature (see, e.g., Dai, 2008). Inaccuracy or uncertainty in making this separation may result in substantial biases in snow accumulation and subsequently in estimates of other fluxes (evaporation, infiltration, groundwater recharge and runoff). If good quality discharge data are available for the model calibration, effects of the threshold value chosen may be assessed and the value can be adjusted through calibration (see more discussion about calibration in Chapter 9). For snow precipitation, a correction factor is also commonly applied to account for snowfall undercatch (see, e.g., Bergström, 1992 and Burek et al., 2013).

8.3 Snowmelt estimation methods

Transformation of snow or ice to liquid water (melting) or to vapour (sublimation) requires energy input. Therefore, a physically based simulation of the snowmelt process involves snowpack energy balance computation, and as such called an energy balance method. An alternative to the energy balance method is a temperature index (or degree-day) method, which is popularly used for its simplicity and least data requirements.

8.3.1 Energy balance method

The energy balance method for snowmelt is similar to the energy balance method for evaporation described in Chapter 3 but with added terms related to snowmelt. The energy balance equation in Chapter 3 (Eq. 3.1) is for a snow fee surface. Additional processes to complete the energy balance equation for snow/ice include energy exchange during melting, freezing and sublimation, and energy added to the snowpack by rainfall. Thus, the energy balance equation for snow or ice surface can be represented by Eq. (8.1) (see, e.g., Anderson, 1976; Oke, 1987; Dingman, 2002; Liston and Elder (2006); DeWalle and Rango, 2008; Andreadis et al., 2009; Cuffey and Paterson, 2010; Sade et al., 2014):

$$\left(R_S^{\downarrow} - R_S^{\uparrow} + R_L^{\downarrow} - R_L^{\uparrow}\right) + G + H + Q_E + Q_r = Q_M + \Delta Q_S \tag{8.1}$$

All the terms in the above equation are energy flux densities with SI unit $\mathrm{J\,m^{-2}\,s^{-1}}$ or $\mathrm{W\,m^{-2}}$. The first four terms on the left-hand side are the shortwave and longwave fluxes (as in Chapter 3), taken downward as positive (input to the snowpack) and upward as negative (outflow from the snowpack). The ground heat flux (G) and sensible heat flux (H) are also described in Chapter 3. Both G and H can be negative or positive depending whether the energy flux is a loss or gain for the snowpack. In Eq. (8.1), energy gain (that is, energy input to the snowpack) is taken positive. The latent heat flux (Q_E) can be evaporation, sublimation or condensation. Evaporation and sublimation fluxes from the snow surface are considered negative (losses from the snowpack) and condensation is considered positive (energy gain for the snowpack). In Eq. (3.1), L_vE was used in place of Q_E, where L_v is the latent heat of vaporization ($\mathrm{J\,kg^{-1}}$) and E is evaporation or condensation as mass flux density ($\mathrm{kg\,m^{-2}\,s^{-1}}$). The latent heat for vaporization or condensation is the same, but different for sublimation, and so it is more convenient to express it as energy flux (Q_E). The latent heat of sublimation (L_s) is taken as the sum of the latent heat of vaporization and latent heat of fusion (L_f), that is, $L_s = L_v + L_f$ (see Table 8.1). Note that in Chapter 3, Eq. (3.1), G, H and E are shown as positive values when the fluxes are outflow from the surface.

The other term on the left-hand side is the energy added by rainfall (Q_r) into the snowpack. Q_r is only required when there is rainfall on the snowpack and is always positive or zero assuming the temperature of rainfall (T_r) $\geq$ the temperature of

Table 8.1 Some thermodynamic properties of water, ice and air.

Properties	Values	References
Specific heat capacity of water (c_w):		
At 15°C	$4186\,\mathrm{J\,kg^{-1}\,°C^{-1}}$	Cutnell and Johnson (2013), p. 361
Specific heat capacity of ice (c_i):		
At 0°C	$2100\,\mathrm{J\,kg^{-1}\,°C^{-1}}$	Oke (1987) p. 44
At −15°C	$2000\,\mathrm{J\,kg^{-1}\,°C^{-1}}$	Cutnell and Johnson (2013), p. 361
Specific heat capacity of air (c_a)	$1010\,\mathrm{J\,kg^{-1}\,°C^{-1}}$	Monteith and Unsworth (2013), p. 375
Specific heat capacity of water vapour (c_v)	$1880\,\mathrm{J\,kg^{-1}\,°C^{-1}}$	Monteith and Unsworth (2013), p. 375
Latent heat of vaporization, L_v:		
At 0°C	$2.501\times10^6\,\mathrm{J\,kg^{-1}}$	Oke (1987), p. 392
At 15°C	$2.465\times10^6\,\mathrm{J\,kg^{-1}}$	Monteith and Unsworth (2013), p. 376
Latent heat of fusion, L_f:		
At 0°C	$0.334\times10^6\,\mathrm{J\,kg^{-1}}$	Oke (1987), p. 392
At −15°C	$0.301\times10^6\,\mathrm{J\,kg^{-1}}$	Oke (1987), p. 392
Latent heat of sublimation, $L_s=L_v+L_f$:		
At 0°C	$2.835\times10^6\,\mathrm{J\,kg^{-1}}$	

snowpack (T_s). The residual of these energy fluxes (that is the net energy flux on the snowpack surface, E_{NET}, see Eq. 8.2) is the energy available for snowmelt (Q_M) and the gain or less in the sensible heat storage of the snowpack (ΔQ_S). Refreezing of liquid water in the snowpack (e.g. the rainwater or water from the snowmelt which has not drained out) can also be included in Q_M, and in that case Q_M (for freezing) is taken negative in Eq. (8.2). Note that sign conventions used in different literature for G, H, Q_E and Q_M may vary.

$$\left.\begin{array}{r}\left(R_S^{\downarrow}-R_S^{\uparrow}+R_L^{\downarrow}-R_L^{\uparrow}\right)+G+H+Q_E+Q_P=E_{NET}\\ E_{NET}=Q_M+\Delta Q_S\end{array}\right\} \tag{8.2}$$

During the period of snow accumulation, E_{NET} is generally negative and Q_M is zero or negligible, so the temperature of the snowpack is generally decreasing ($\Delta Q_S<0$). During the melting period, the snowpack temperature needs to be raised to 0°C before melting starts. This is called the warming phase of the melting period when the snowpack temperature starts rising but still below the melting temperature. During the warming phase, E_{NET} is generally positive but Q_M is still zero, and so the temperature of the snowpack increases ($\Delta Q_S>0$) proportional to E_{NET}. Note that, if temperature on the surface is at melting point but colder below the surface, the meltwater from the surface may percolate into the snowpack and refreeze. In that case Q_M is negative (Eqs. 8.1 and 8.2) and the energy is input to the snowpack. The change in average temperature of the snowpack (ΔT_S) over a time period Δt can be related to the net energy flux by the following equation

$$\Delta T_s = \frac{\Delta Q_S \Delta t}{c_i \rho_w \mathrm{SWE}} \tag{8.3}$$

where Δt is in s (dimension T), ΔT_S is in °C, c_i is the heat capacity of ice ($\mathrm{J\,kg^{-1}\,°C^{-1}}$), ρ_w is the density of water ($\mathrm{kg\,m^{-3}}$) and SWE is snow water equivalent (m), which is the equivalent water depth of the total snowpack depth, given by

$$\mathrm{SWE} = \frac{\rho_s}{\rho_w} h_s \tag{8.4}$$

where ρ_s is the snowpack density ($\mathrm{kg\,m^{-3}}$) and h_s is the snowpack depth (m). Note that the snow density and the estimation of SWE include any water that may be present in the snowpack as well as the ice crystals (DeWalle and Rango, 2008). Eq. (8.3) is also used to estimate the cold content of the snowpack, which is defined as the amount of energy required to raise its temperature to the melting point (see, e.g. Dingman, 2002; Jennings et al., 2018). Thus, the cold content (Q_{cc}) ($\mathrm{J\,m^{-2}}$) is given by

$$Q_{cc} = -c_i \rho_w \mathrm{SWE}(T_s - T_m) \tag{8.5}$$

where T_m and T_s are the melting temperature and snowpack temperature, respectively.

Once the snowpack temperature is raised to the melting point, the snowpack is assumed to maintain more or less isothermal condition, so ΔQ_S is assumed to be zero. Then the residual energy flux equals the energy used for snowmelt (i.e. $Q_M = E_{NET}$), and the melt can be computed as

$$M_s = \frac{Q_M}{\rho_w L_f} \tag{8.6}$$

where M_s is the snowmelt rate ($\mathrm{m\,s^{-1}}$).

The main difficulty in applying the energy balance method resides on accurately estimating each of the terms in Eq. (8.1). In the absence of measured data, the shortwave and longwave radiations can be estimated using the same approach discussed in Chapter 3 and Eqs. (3.4)–(3.7). The ground heat flux can also be estimated using Eq. (3.9) if the soil thermal conductivity is known. For snowpack energy balance, it is usually found a small quantity compared to other flux terms and sometimes a small positive constant value is assumed (Warscher et al., 2013; DeWalle and Rango, 2008).

The sensible heat transfer is commonly expressed with the temperature difference between the air in the atmosphere and at the snowpack surface, and the aerodynamic resistance factor (r_a) (Eq. 8.7). Similarly, the latent heat transfer is expressed with the vapour pressure gradient and the aerodynamic resistance factor (Eq. 8.8).

$$H = c_a \rho_a \frac{(T_a - T_s)}{r_a} \tag{8.7}$$

$$Q_E = L_{v,s}\rho_w \frac{0.622}{P_a}\frac{(e_a - e_s)}{r_a} \tag{8.8}$$

In Eq. (8.7), r_a is in s m^{-1} (dimension T L^{-1}), c_a is the heat capacity of air (J kg^{-1} °C^{-1}), ρ_a is air density (kg m^{-3}), and T_s is the snow surface temperature and T_a is the air temperature (°C) at a height z from the ground. In Eq. (8.8), P_a is the atmospheric pressure (kPa), $L_{v,s}$ is latent heat of vaporization or sublimation and e_a and e_s are vapour pressures (kPa) in the air at the snow surface and at a height z from the ground, respectively. Note also that in Eq. (8.7), $(T_a - T_s)$ is used instead of $(T_s - T_a)$ and in Eq. (8.8), $(e_a - e_s)$ is used instead of $(e_s - e_a)$, so that H and Q_E have the correct sign (negative/positive) according to Eqs. (8.1) and (8.2). To estimate the latent heat flux for the snow surface, an additional difficulty arises to separate between evaporation and sublimation. One approach often used is by checking whether the snowpack contains meltwater. If the snowpack has liquid (water) content, the latent heat flux loss is considered as evaporation and if the snowpack temperature is below freezing then considered as sublimation (see, e.g., Dingman, 2002; DeWalle and Rango, 2008).

The aerodynamic resistance is defined as (see, e.g., Andreadis et al., 2009)

$$r_a = \frac{\ln\left(\frac{z - d_s}{z_0}\right)^2}{k^2 u_z} \tag{8.9}$$

where z is the height (m) from the ground where u_z, T_a and e_a are measured, d_s is the snow depth, u_z is the wind velocity (m s^{-1}) at height z, z_0 is the surface roughness height and k is von Karmin's constant = 0.4. Note that if the height z is measured from the surface of the snowpack, d_s is replaced by z_d, which is called the zero-displacement height and usually considered negligible (see, e.g. Dingman, 2002). The range of variation of z_0 is quite large, and the values reported in the literature usually vary between 0.001 m to 0.005 m for average snowpack surface condition. However, the range can be much larger for very smooth to very rough snowpack surfaces (DeWalle and Rango, 2008). The application of Eqs. (8.6)–(8.8) for sensible and latent heat fluxes is further complicated due to the stability condition of the atmospheric boundary layer. More discussions on this and corrections that may be applied can be found in Anderson (1976), Andreadis et al. (2009), Dingman (2002) and DeWalle and Rango (2008).

The advective energy flux due to rainfall ('rain-on-snow') can be estimated using Eq. (8.10) or Eq. (8.11) depending on whether the snowpack temperature is at melting point (usually taken as 0°C) or at below freezing.

$$Q_r = c_w \rho_w r (T_r - T_m) \tag{8.10}$$

$$Q_r = c_w \rho_w r (T_r - T_m) + \rho_w r L_f \tag{8.11}$$

where c_w is the heat capacity of water (J kg^{-1} °C^{-1}), r is the rainfall rate (m s^{-1}) and T_r is the temperature of rainwater (°C). In Eq. (8.11), the second term represents the energy released by the rainwater when it freezes on the snowpack.

Examples of computing each of these terms can be found in Dingman (2002) and DeWalle and Rango (2008).

8.3.2 Temperature index method

The energy balance approach for snowmelt is demanding in terms of data requirements, and adequate data are often unavailable for a catchment scale modelling. As a result, many hydrological models use a simple empirical approach called a temperature index or degree-day method. The concept and equation of the temperature index method are simple. It defines an empirical relationship to estimate the amount of snowmelt as a result of per degree rise in temperature above the freezing/melting threshold temperature, that is

$$M_s = \begin{cases} f_{dd}(T_a - T_m) & \text{if } T_a \geq T_m \\ 0 & \text{if } T_a < T_m \end{cases} \quad (8.12)$$

where M_s is the snowmelt rate (mm day^{-1}), f_{dd} is the degree-day (or temperature index) melt factor, in short 'degree-day factor' (mm °C^{-1} day^{-1}), and T_a and T_m are the daily average air temperature and melting temperature threshold. T_m is usually set to 0, but slightly different values (±) are also used. How the T_a is estimated may vary, but the average of daily minimum and maximum temperature is commonly used.

There are several factors that affect the value of the degree-day factor, e.g. weather conditions, latitude, topography, land cover, and time of the year. Consequently, its values reported in the literature vary significantly from about 1 to 8 mm °C^{-1} day^{-1} (DeWalle and Rango, 2008). See Hock (2003) for a comprehensive review of the degree-day factor, particularly on mountainous areas and glaciated sites. Seasonal variation is commonly applied to the degree-day factor, typically using a sinusoidal variation over the melt period of year, e.g., used in LISFLOOD (Burek et al., 2013) and SWAT (Neitsch et al., 2011).

In catchment modelling, the degree-day factor is essentially treated as a calibration parameter. The range of values reported in the literature, ideally from hydrologically similar type of catchments, provides a good starting value/range for calibration and for uncertainty estimation.

8.4 Snowpack mass balance and snowmelt runoff

In Section 8.3, two methods of snowmelt estimation are described, one based on the energy balance approach and one temperature index approach. When the meltwater content exceeds the water folding capacity of the snowpack, the meltwater leaves the snowpack through infiltration into the soil or as direct surface runoff, and thus the snowpack mass changes. Snowpack also loses mass by evaporation and sublimation (discussed in Section 8.3.1). Precipitation and condensation on the snow surface result in snowpack mass gain (or accumulation). This change in snowpack mass (in snow water equivalent) can be represented as

$$\frac{\partial \text{SWE}}{\partial t} = P_s - E_{e,s} - M_{s,\text{out}} \tag{8.13}$$

where P_s is the precipitation rate on the snowpack, $E_{e,s}$ is the evaporation or sublimation rate and $M_{s,out}$ is the meltwater outflow rate (which is the meltwater exited from the snowpack as infiltration into the soil or as direct runoff). All three terms (on the right-hand side) have the same unit, that is [LT^{-1}]. Note that interception and throughfall are not shown in the equation. The condensation (mass gain) is usually considered small quantity and also not shown in this equation.

The meltwater outflow is related to the snowmelt (M_s) computed by the snow energy balance (Eq. 8.5) or the temperature index method (Eq. 8.12) and the liquid water content (LWC) in the snowpack. If the snowpack has already reached the maximum LWC the snowpack can hold (LWC_{max}), then all the additional snowmelt becomes outflow, that is $M_{s,out} = M_s$. If the maximum water holding capacity has not been reached yet, the snowmelt is first used to satisfy the LWC_{max}.

Note that Eq. (8.13) assumes a single layer with uniform snow density over the snowpack depth. If the density difference as well as temperature difference over the snow profile are to be included, vertically distributed representations are required for Eq. (8.13) as well as the melt estimation equations presented in Sections 8.3.1 and 8.3.2.

Furthermore, Eq. (8.13) does not include space dimension, which means that it assumes an area of uniform snow cover thickness and other parameters including weather parameters. Such an estimation is also referred to as a 'point estimate' of snowmelt. When it is applied on a catchment scale, the spatial variabilities need to be taken into account. These includes the variations in the snow cover area and snow cover depth, variation in weather input parameters and variation in the land cover within the spatial computational unit. The computational units are grid-cells in case of a spatially distributed model structure, and subcatchments in case of a semidistributed type model structure. In some models, hydrologic response units are also used, which are subunits within a subcatchment or grid-cell.

Additionally, transport of snow mass due to wind or gravity forces also contributes to the dynamic variation of snow cover area in a catchment. Many hydrological models allow to apply elevation bands (that is dividing a subcatchment or grid-cell into areas of different elevation ranges) with lapse rates for weather parameters, particularly temperature. This usually results in a significant improvement in snowmelt simulation. The wind and gravity driven transport of snow is also incorporated in some models. The latter one is particularly important in mountainous catchments with steep terrain (Warscher et al., 2013).

8.5 How different catchment models treat snowmelt runoff?

Methods and approaches used in 17 different catchment models for snow and ice melt runoff are reviewed and briefly described here (Table 8.2). The selected models are among the widely used models, but the list is not exhaustive, and the purpose of

Table 8.2 Snowmelt methods used in various hydrological models.

Modelling software	Method available	Additional information
BROOK90	Snowmelt is based on the temperature index method. It also takes into account snowpack's cold content and snow temperature. Canopy interception of snowfall and snow evaporation are also included. (Federer, 2002). See also Dingman (2002).	
CASC2D	It does not have a snowmelt model. Ogden (2001), see also Downer et al. (2002).	
CHARM (also WATFLOOD)	Snowmelt is based on the temperature index method, and it also estimates energy deficit of snow cover to distinguish between melt period and nonmelt period, that is when the snow cover is not "ripe" for melting. It also has a 'temperature and radiation' index method, whereby in addition to the temperature index based melt, melt from the surface radiation balance is added (Kouwen, 2018).	CHARM: Canadian Hydrological and Routing Model
Flo2D	It primarily simulates flow routing and flood inundation and does not include snow input and snowmelt (FLO-2D, 2003).	
HBV (and HBV-light)	Snowmelt is based on the temperature index (degree-day) method. The meltwater remains in the snowpack until it exceeds a specified threshold value to account for the liquid water holding capacity of snow, and this delays the snowmelt runoff. The meltwater in the snowpack is allowed to refreeze also based on the temperature index approach. Rainfall or snowfall differentiation from input precipitation is based on a threshold temperature, with a possibility to adjust the snowfall input through a correction	

Continued

Table 8.2 Snowmelt methods used in various hydrological models—cont'd

Modelling software	Method available	Additional information
	factor (Bergström, 1992; Seibert, 2005). See also Lindström et al. (1997)	
HEC-HMS	Snowmelt is based on the temperature index method. The cold content of the snowpack is also used and estimated based on the cold content Antecedent Temperature Index (ATI). The ATI is updated based on the air temperature of the current time step and a factor called the ATI cold-rate coefficient. Elevation bands and temperature lapse rates can be applied to account for elevation distributed snowmelt (HEC-HMS, 2005, 2010, 2020).	Note that the snow module is added in HEC-HMS from version 3.0.0 (HEC-HMS, 2005), and so the Technical Reference Manual (Feldman, 2000) does not include the snowmelt module description.
LISFLOOD	Snowmelt is based on the temperature index method but with an additional coefficient for seasonal variation of the melt rate and additional correction for snowmelt on rainy days (increased melt rate with increased rainfall intensity). In the high altitudes where accumulated snow does not melt completely year by year, the residual snow accumulation is treated as ice and the ice melt, in addition to the snowmelt, is applied during the summer months. Although LISFLOOD is a distributed model (grid-based structure), snowmelt is applied in three elevation zones within a grid cell. This is to account for possible subgrid variation of elevation and temperature when the model is applied with a coarse model resolution. Separation of rain or snow from precipitation input is based on the threshold temperature and a snow correction factor is also allowed to correct for possible	

Table 8.2 Snowmelt methods used in various hydrological models—cont'd

Modelling software	Method available	Additional information
	snow undercatch (Burek et al., 2013).	
MIKE SHE	Snowmelt is based on a temperature index or degree-day ("modified") approach, but the melt is estimated separately for temperature, incoming solar radiation and added energy due to liquid rain. Snowpack is divided into dry and wet snow. Snowmelt is first treated as wet snow, and snowmelt runoff is allowed only when the wet / dry snow ratio exceeds a specified threshold. Refreezing is also considered by increasing the dry snow fraction. Precipitation and temperature are both corrected for elevation using lapse rates. Snow evaporation from wet snow and sublimation from dry snow are also included. It also allows varying the snow coverage area within a grid-cell by specifying a minimum snowpack depth above which the entire cell area is covered with snow (DHI, 2017a, 2017b).	See also https://manuals.mikepoweredbydhi.help/2017/MIKE_SHE.htm.
NAM	The snowmelt module is very much similar to that of MIKE SHE. Snowmelt is based on simple temperature index method with a possibility for the extended option for additional radiation melt and melt due to added energy from liquid rain (see above description for MIKE SHE). Subcatchments can be divided into altitude bands for distributed snowmelt with corrected precipitation and temperature based on specified lapse rates. Seasonal variation of the snowmelt coefficient can also be applied (DHI, 2017c).	Note that the description presented here is based on the NAM model used in MIKE 11, the river modelling software of DHI (https://manuals.mikepoweredbydhi.help/2017/MIKE_11.htm.).

Continued

Table 8.2 Snowmelt methods used in various hydrological models—cont'd

Modelling software	Method available	Additional information
PCR-GLOBWB	The separation of rain or snow precipitation is based on a temperature threshold. Snowmelt is based on the temperature index method and follows the same procedure as the snow module of HBV model (Van Beek and Bierkens, 2009). See above in this table about the HBV model.	This description is based on PCR-GLOBWB 1.0, and currently version 2.0 is also available https://globalhydrology.nl/research/models/pcr-globwb-2-0/), but the key approach/method for the snow module is the same. See also Sutanudjaja et al. (2018)
PRMS (Ver. 4) also used in GSFLOW	The snowmelt model is based on the energy and mass balance approach and computes at the level of hydrologic response unit. It also simulates albedo (fresh snow generall has higher albedo than old snow), density and snow cover area (Markstrom et al., 2015).	PRMS: Precipitation Runoff Modelling System (Markstrom et al., 2015). GSFLOW: Coupled Ground-water and Surface-water Flow Model (Markstrom et al., 2008).
RRI model	It does not have a snowmelt module until version 1.4.2 (2017) referred to in this review (Sayama, 2017).	
SWAT	Snowmelt is based on the temperature index method and considers both the snowpack temperature and air temperature. The snowpack temperature is estimated for the current day average air temperature and previous day's snowpack temperature using a weighting factor called a "lag" (between 0 and 1). Seasonal variation of the melt factor is also used. Elevation bands with the temperature lapse rate can be applied to account for snowmelt variation within subcatchments. Variation in the snow cover area for a given depth of snow is modelled defining an "areal depletion curve" with a threshold snowpack depth (in snow water equivalent) above which snow cover occupies the entire subcatchment (Neitsch et al., 2011).	

Table 8.2 Snowmelt methods used in various hydrological models—cont'd

Modelling software	Method available	Additional information
VIC model	The snow model is based on the energy and mass balance approach. The energy balance equation is based on Andreadis et al. (2009) and some of the procedures are based on Anderson (1976). To account for subgrid variation of snowmelt, grid-cells are partitioned into elevation bands and each elevation band into different land covers (Gao et al., 2010).	VIC (Variable Infiltration Capacity). See also the VIC version 5 webpage: https://vic.readthedocs.io/en/master/
WaSiM	It has a range of options for snowmelt, such as the simple temperature index method, temperature and wind index method, "combination approach" based on Anderson (1973) and energy balance method based on Warscher et al. (2013). It also applies wind driven redistribution of snow, which works like a snow precipitation correction, and gravity driven redistribution of snow to account for the transport of accumulated snow mass on steep sloped terrain. Snow sublimation is also calculated separately as part of the energy balance computation of snowmelt/sublimation (Schulla, 2021).	See also the version updates on the WaSiM webpage: http://www.wasim.ch/en/the_model/dev_details.htm
Xinanjiang Model	It does not have a snowmelt model in the versions referred to in this review (Zhao, 1992; Yao et al., 2012; Fang et al., 2017).	

presenting the table is not for giving the author's judgement about the models. It is simply intended to describe the key concepts of the methods and approaches used in the models. The descriptions attempt to capture the main essence or features but may not be detailed enough to provide complete details of the methods, for which the referenced literature should be consulted. Most models have different versions and some methods used between versions may vary. The descriptions presented here are based on the references (mostly the reference manual of the corresponding models) cited in the table.

References

Anderson, E.A., 1973. National Weather Service River Forecast System—Snow Accumulation and Ablation Model. NOAA Technical Memorandum NWS HYDRO-17, 217 pp.

Anderson, E.A., 1976. A point energy and mass balance model of a snow cover. NOAA Technical Report NWS 19, 150 pp.

Andreadis, K.M., Storck, P., Lettenmaier, D.P., 2009. Modeling snow accumulation and ablation processes in forested environments. Water Resour. Res. 45, W05429. https:/doi.org/10.1029/2008WR007042.

Bergström, S., 1992. The HBV model—its structure and applications. SMHI RH No. 4. Norrköping.

Burek, P., Van der Knijff, J., De Roo, A., 2013. LISFLOOD distributed water balance and flood simulation model, revised user manual. JRC Technical Report 'EUR 26162 EN'.

Cuffey, K.M., Paterson, W.S.B., 2010. The Physics of Glaciers, fourth ed. Elsevier.

Cutnell, J.D., Johnson, K.W., 2013. Introduction to Physics, 9th. John Wiley and Sons, Inc, Singapore.

Dai, A., 2008. Temperature and pressure dependence of the rain-snow phase transition over land and ocean. Geophys. Res. Lett. 35. https:/doi.org/10.1029/2008gl033295, 112802.

DeWalle, D., Rango, A., 2008. Principles of Snow Hydrology. Cambridge University Press.

DHI (2017a). MIKE SHE Volume 2: Reference Guide. DHI, Denmark. https:/manuals.mikepoweredbydhi.help/2017/Water_Resources/MIKE_SHE_Printed_V2.pdf; Accessed on 31 July, 2021.

DHI (2017b). MIKE SHE Volume 1: User Guide. DHI, Denmark. https:/manuals.mikepoweredbydhi.help/2017/Water_Resources/MIKE_SHE_Printed_V1.pdf; Accessed on 31 July, 2021.

DHI (2017c). MIKE 11 A Modelling System fors Rivers and Channels, Reference Manual. DHI, Denmark. https:/manuals.mikepoweredbydhi.help/2017/Water_Resources/Mike_11_ref.pdf; Accessed on 31 July, 2021.

Dingman, S.L., 2002. Physical Hydrology, second ed. Prentice Hall, New Jersey.

Downer, C.W., Ogden, F.L., Martin, W.D., Harmon, R.S., 2002. Theory, development, and applicability of the surface water hydrologic model CASC2D. Hydrol. Process. 16, 255–275.

Fang, Y.-H., Zhang, X., Corbari, C., Mancini, M., Niu, G.-Y., Zeng, W., 2017. Improving the Xin'anjiang hydrological model based on mass-energy balance. Hydrol. Earth Syst. Sci. 21, 3359–3375. https:/doi.org/10.5194/hess-21-3359-2017.

Federer, C.A., 2002. BROOK 90: a simulation model for evaporation, soil water, and streamflow. http:/www.ecoshift.net/brook/brook90.htm.

Feldman, A.D. (Ed.), 2000. Hydrologic Modelling System HEC-HMS Technical Reference Manual. US Army Corps of Engineers, Hydrologic Engineering Centre, Washington, DC.

FLO-2D, 2003. FLO-2D User Manual. Nutrioso.

Gao, H., Tang, Q., Shi, X., Zhu, C., Bohn, T.J., Su, F., Sheffield, J., Pan, M., Lettenmaier, D.P., Wood, E.F., 2010. Water Budget Record from Variable Infiltration Capacity (VIC) Model. In: Algorithm Theoretical Basis Document for Terrestrial Water Cycle Data Records. UNSPECIFIED. Available for download from https:/eprints.lancs.ac.uk/id/eprint/89407.

HEC-HMS, 2005. Hydrologic Modelling System, HEC-HMS User's Manual Version 3.0.0. U. S. Armey Corps of Engineers, Hydrologic Engineering Centre, Davis, DA.

HEC-HMS, 2010. Hydrologic Modelling System, HEC-HMS Release Notes Version 3.5. U.S. Armey Corps of Engineers, Hydrologic Engineering Centre, Davis, DA.

HEC-HMS (2020). HEC-HMS User's Manual, Version 4.8. https:/www.hec.usace.army.mil/confluence/hmsdocs/hmsum/4.8; last accessed on 25 Septemer, 2021.
Hock, W., 2003. Temperature index melt modelling in mountain areas. J. Hydrol. 282, 104–115. https:/doi.org/10.1016/S0022-1694(03)00257-9.
Jennings, K.S., Kittel, T.G.F., Molotch, N.P., 2018. Observations and simulations of the seasonal evolution of snowpack cold content and its relation to snowmelt and the snowpack energy budget. Cryosphere 12, 1595–1614. https:/doi.org/10.5194/tc-12-1595-2018.
Kouwen, N., 2018. WATFLOOD/CHARM Canadian Hydrological and Routing Model. University of Waterloo, Canada. http:/www.civil.uwaterloo.ca/watflood/index.htm.
Lindström, G., Johansson, B., Persson, M., Gardelin, M., Bergström, S., 1997. Development and test of the distributed HBV-96 hydrological model. J. Hydrol. 201, 272–288. https:/doi.org/10.1016/S0022-1694(97)00041-3.
Liston, G.E., Elder, K., 2006. A distributed snow-evolution modeling system (SnowModel). J. Hydrometeorol. 7 (6), 1259–1276. https:/doi.org/10.1175/JHM548.1.
Markstrom, S.L., Niswonger, R.G., Regan, R.S., Prudic, D.E., Barlow, P.M., 2008. GSFLOW-Coupled Ground-water and Surface-water FLOW model based on the integration of the Precipitation-Runoff Modeling System (PRMS) and the Modular Ground-Water Flow Model (MODFLOW-2005). U.S. Geological Survey Techniques and Methods 6-D1, 240 p.
Markstrom, S.L., Regan, R.S., Hay, L.E., Viger, R.J., Webb, R.M.T., Payn, R.A., LaFontaine, J.H., 2015. PRMS-IV, the Precipitation-Runoff Modeling System, Version 4, Techniques and Methods 6–B7. U.S. Geological Survey, Reston, Virginia.
Monteith, J.L., Unsworth, M.H., 2013. Principles of Environmental Physics, 4th. Academic Press.
Neitsch, S.L., Arnold, J.G., Kiniry, J.R., Williams, J.R. (2011). Soil and Water Assessment Tool Theoretical Documentation Version 2009. Texas Water Resources Institute. Available electronically from https:/hdl.handle.net/1969.1/128050.
Ogden, F.L. (1998, rev 2001). A Brief Description of the Hydrologic Model CASC2D. Univ. Connecticut.
Oke, T.R., 1987. Boundary Layer Climate, second ed. Routledge.
Sade, R., Rimmer, A., Litaor, M.I., Shamir, E., Furman, A., 2014. Snow surface energy and mass balance in a warm temperate climate mountain. J. Hydrol. 519, 848–862. https:/doi.org/10.1016/j.jhydrol.2014.07.048.
Sayama, T., 2017. Rainfall-Runoff Inundation (RRI) Model, Version 1.4.2. International Centre for Water Hazard and Risk Management (ICHARM) and Public Works Research Institute (PWRI), Japan.
Schulla, J., 2021. Model description WaSiM, Version 10.06.00. Hydrology Software Consulting J. Schulla, Zurich. http:/www.wasim.ch/downloads/doku/wasim/wasim_2021_en.pdf.
Seibert, J., 2005. HBV Light Version 2 User's Manual. Department of Physical Geography and Quaternary Geology, Stockholm University.
Sutanudjaja, E.H., van Beek, R., Wanders, N., Wada, Y., Bosmans, J.H.C., Drost, N., van der Ent, R.J., de Graaf, I.E.M., Hoch, J.M., de Jong, K., Karssenberg, D., López López, P., Peßenteiner, S., Schmitz, O., Straatsma, M.W., Vannametee, E., Wisser, D., Bierkens, M.F.P., 2018. PCR-GLOBWB 2: a 5 arcmin global hydrological and water resources model. Geosci. Model Dev. 11, 2429–2453. https:/doi.org/10.5194/gmd-11-2429-2018.
Van Beek, L.P.H., Bierkens, M.F.P., 2009. The Global Hydrological Model PCR-GLOBWB: Conceptualization, Parameterization and Verification, Report. Department of Physical

Geography, Utrecht University, Utrecht, The Netherlands. http:/vanbeek.geo.uu.nl/suppinfo/vanbeekbierkens2009.pdf.
Warscher, M., Strasser, U., Kraller, G., Marke, T., Franz, H., Kunstmann, H., 2013. Performance of complex snow coverdescriptions in a distributed hydrological model system: a case study for the high alpine terrain of the Berchtesgaden Alps. Water Resour. Res. 49, 2619–2637. https:/doi.org/10.1002/wrcr.20219.
Yao, C., Li, Z., Yu, Z., Zhang, K., 2012. A priori parameter estimates for a distributed, grid-based Xinanjiang model using geographically based information. J. Hydrol. 468–469 (2012), 47–62. https:/doi.org/10.1016/j.jhydrol.2012.08.025.
Zhao, R.-J., 1992. The Xinanjiang model applied in China. J. Hydrol. 135, 371–381.

CHAPTER

Model components integration, model calibration and uncertainty

9

9.1 Integration of model components

A comprehensive catchment model comprises of components representing various 'precipitation to runoff' processes discussed in Chapters 3–8. These components are connected to each other directly or through other components, and the connection can be one way or two way (Fig. 9.1). The direction of the arrow indicated in the connection is based on the direction of the flux movement between the two components. For example, evaporation and transpiration take place from the soil water, so the flux direction is from the Unsaturated Zone component to the Evaporation component, which is represented by connection (1). In the diagram, the Evaporation and Interception are shown as one component for simplicity although they are two separate components. The connection (2) with the arrow towards the Unsaturated Zone is to represent the throughfall from the Interception component. In the similar way, the Evaporation and Interception component is connected to the Snowmelt and River Flow Routing components. However, not all the connections are represented in all models. For example, direct evaporation from river water may be considered negligible compared to the evaporation from the entire catchment area. However, in a large river system with a large water surface area, evaporation loss can be nonnegligible, so the connection should be included. The Evaporation and Interception components are not directly connected to the Groundwater Flow and Surface Flow Routing components, but connected through the Unsaturated Zone. In fact, in some catchments, some plants may take water from the Saturated Zone directly for transpiration. In such a case the Evaporation component is also directly connected to the Groundwater Flow component, which is not shown in the diagram. As we can see in the figure, the Unsaturated Zone is connected to every other component of the catchment model. Note that a simplified catchment model may not have all the components, and depending on the hydrological conditions one or more components may not be necessary. For example, some models may not have the interception component (explicitly) or river routing component, while the snowmelt component is not necessary if the catchment is only rainfed. In cases where a component is not modelled, its effect is either combined with other components or considered negligible.

Implementing all the connections in the model is complex. Most models apply some kind of simplification to mathematically represent the connections. For example, an order is usually defined to satisfy the evaporation demand from interception storage and soil water, which is to simplify the connection of the evaporation

Catchment Hydrological Modelling. https://doi.org/10.1016/B978-0-12-818337-3.00009-X

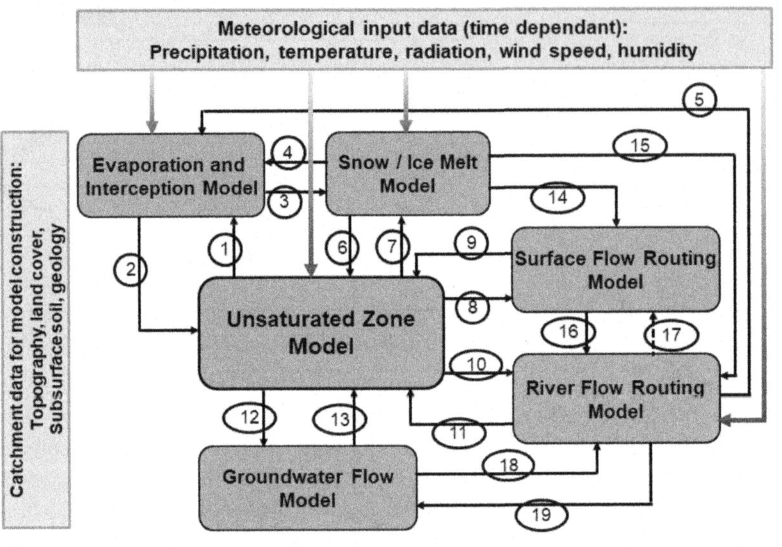

FIG. 9.1

Schematic diagram of various catchment hydrological model components and their connections. Connections indicated by the numbers represent: (1) Water extraction from the soil by direct evaporation or plant transpiration, (2) Part of the intercepted rainwater throughfall, (3) Throughfall of intercepted snow, (4) Evaporation and sublimation from the snowpack, (5) Evaporation from river water (open water evaporation), (6) Infiltration input from the meltwater outflow, (7) Refreezing of the meltwater and ground heat flux exchange between snowpack and Unsaturated Zone soil, (8) Surface flow input to the Surface Flow Routing, (9) *Re*-infiltration of the surface flow before reaching the streams, (10) Interflow (Q_{if}) input to the River Flow Routing, (11) Infiltration from river beds to the Unsaturated Zone in parts of the river reach not linked with the Saturated Zone, (12) Recharge input to the Groundwater Flow Model, (13) Moisture transport through capillary rise, (14) Direct runoff from the meltwater output to the Surface Flow Routing, (15) Direct snowmelt input to the river, if exists, (16) Surface runoff (Q_{sf}) input to the river, (17) Overbank flow to the flood plain which may become part of surface flow, (18) Groundwater input to the river (Q_{gw}), and (19) Recharge to the groundwater from the river. Note that not all hydrological models represent all the connections shown here.

component with the interception and unsaturated zone components. Similarly, to estimate the recharge from river to groundwater the hydraulic heads may be taken from the previous time step, which is like using an explicit numerical scheme to simplify the hydraulic routing of surface or river flow discussed in Chapters 5 and 7.

In the figure, the model input data are indicated in two separate blocks, one for the weather related input data which are time varying inputs and one for representing the physical characteristics of the catchment. The latter data sets are primarily the DEM (digital elevation model) and soil and land cover data, which are usually considered constant throughout the model simulation period, with the exception of some types of land cover. In particular the vegetation land cover has seasonal variation, which is usually kept constant year to year. Note that one of the applications of a catchment model is to study the impact of land use change in catchment hydrology and water resources, and obviously for such applications the changes in the land use need to be taken into account. See Chapter 2 for more on data types.

9.2 How good is my model?

'How good is my model?' is a basic question that every modeller would like to have an answer. It is however not an easy question and the answer may be different by different people. One standard practice to try and answer this question is by comparing the model results against observation data. Graphical plots of such a comparison tell a lot about how two data series (simulated by the model and observed) match each other. To quantitatively indicate the "goodness-of-fit" between them, several metrics, called performance metrics, can be used. However, the performance metrics do not give all the answers and the interpretation of the metrics can still be a subjective matter. In general, use of two or three different metrics gives a lot better indication of the model performance. But there are several things that are worth knowing before we conclude how good is a model, which are discussed in the following sections.

9.2.1 Compare against what?

Catchment hydrological models are mostly used for precipitation to runoff simulation, and the runoff or more specifically the river discharge is the primary model output. So the comparison is almost always made between observed and model simulated discharge time series. Note that observed or measured discharge is sometimes a misleading term and can be a major source of uncertainty. We discuss about uncertainty in Section 9.4, so for the discussion here let us assume the observed discharge is free of errors. Apart from this there are at least two major issues. First, discharge is not the only thing the catchment model computes and also not the only concerned output variable. Evaporation is a significant component of the catchment water balance, but good model performance in simulating discharge does not automatically mean equally good performance in simulating evaporation. Even for the discharge, as we discussed in Chapters 4, 5, and 6 that river discharge comprises of surface runoff, interflow and groundwater flow (although not all the time), but the observed discharge that may be available for comparison is almost always a total discharge. Similarly, what about changes in groundwater and surface storages (e.g. lakes and reservoirs), and changes in snow/ice deposits? These changes are collectively referred to as the change in storage over time which is also an important unknown in catchment modelling. In some applications of a catchment model, understanding the changes in groundwater may be the key objective, e.g. to study the impact of land use change on surface water/groundwater resources in the catchment. If the purpose of the model application is for assessing the water use for agriculture, quantification of evaporation becomes a key target. If the model is aimed at applying for high flood discharge estimation, evaporation may be less important as during high rainfall events, contribution of evaporation relative to rainfall is usually small. On the other hand, for drought assessment, which is a longer period event than floods, evaporation is as important.

Second, discharge is a point measurement, so what about the spatial variability of precipitation–runoff response in the catchment? Discharge data is usually not readily available and a lot scarcer than rainfall data. We are unlikely to have sufficient observed discharge data to spatially evaluate the model performance on discharge simulation across the catchment. So how can we assess the model performance spatially?

The reason why hydrological models are always calibrated only with discharge data, with the exception of very few, is because of the lack of data itself. Good news is that it has started to change in particular with the advancement of remote sensing based observations and big data science (see e.g. Maskey, 2019; Sirisena et al., 2020). Satellite remote sensing based evaporation and snow cover data are more and more commonly available these days, although their quality and reliability need to be assessed too (see, e.g. Miralles et al., 2011; Mu et al., 2011; Trambauer et al., 2014; Hall et al., 2002; Maskey et al., 2011; Notarnicola, 2020). Now is probably the right time that we move ahead from using only the discharge data as a measure for assessing a hydrological model and include more of these other emerging data where available to make the assessment more comprehensive. By bringing remote sensing based data such as evaporation and snow cover, we are expanding model assessments from point measurement data (river discharge) to spatial data. It is not that these emerging data are perfect and we can treat them with the similar confidence as in situ observation data, but they certainly bring additional information and open up a new paradigm for calibrating hydrological models spatially. How we bring these different types of data with different levels of confidence or uncertainty into the calibration framework is a different question and also an emerging research question.

9.2.2 Model performance metrics

Another issue is related to the type of metrics we use for model performance evaluation. Different types of metrics are available, but few of them dominate in most hydrological modelling literature. The Nash-Sutcliff efficiency (NSE) is among the most widely used metrics in hydrology. The coefficient of determination (r^2), mean error or bias (ME) or percent bias (PBIAS), root-mean squared error (RMSE) and more recently Kling-Gupta efficiency (KGE) are also widely used. Mathematical expressions of these and several commonly used metrics are presented in Table 9.1. Obviously, each indicator has different meaning, and the use of one metric is often inadequate to usefully interpret the model performance. That is why two or more metrics are commonly used together.

Table 9.1 Model performance evaluation metrics. Some other metrics that are not presented here can be found in Gupta et al. (1998), Hall (2001) and Chadalawada and Babovic (2019).

Metrics	Formula
Mean error (average bias)	$ME = \frac{1}{n}\sum_{i=1}^{n}(y_{s,i} - y_{o,i})$
Mean absolute error (MAE)	$MAE = \frac{1}{n}\sum_{i=1}^{n}\lvert y_{s,i} - y_{o,i}\rvert$
Percent bias (PBAIS)	$PBIAS = \frac{ME}{\bar{y}_o} \times 100$ where $\bar{y}_o = \frac{1}{n}\sum_{i=1}^{n} y_{o,i}$
Mean squared error (MSE)	$MSE = \frac{1}{n}\sum_{i=1}^{n}(y_{s,i} - y_{o,i})^2$
Root mean squared error (RMSE)	$RMSE = \sqrt{\frac{1}{n}\sum_{i=1}^{n}(y_{s,i} - y_{o,i})^2}$
Correlation coefficient (r)	$r = \frac{\frac{1}{n}\sum_{i=1}^{n}(y_{o,i}y_{s,i}) - \bar{y}_o\bar{y}_s}{\sqrt{\frac{1}{n}\sum_{i=1}^{n}y_{o,i}^2 - (\bar{y}_o)^2}\sqrt{\frac{1}{n}\sum_{i=1}^{n}y_{s,i}^2 - (\bar{y}_s)^2}}$
Coefficient of determination (CoD, r^2, or R^2)	$CoD = r^2$
Nash-Sutcliffe efficiency (NSE)	$NSE = 1 - \frac{\sum_{i=1}^{n}(y_{s,i} - y_{o,i})^2}{\sum_{i=1}^{n}(y_{o,i} - \bar{y}_o)^2} = 1 - \frac{MSE}{\sigma_o^2}$
Modified Nash-Sutcliffe efficiency	$MNSE = 1 - \frac{\sum_{i=1}^{n}\lvert y_{s,i} - y_{o,i}\rvert^p}{\sum_{i=1}^{n}\lvert y_{o,i} - \bar{y}_o\rvert^p}$ where $1 \le p \le 2$; $p = 1$ is commonly used.
Kling-Gupta efficiency (KGE)	$KGE = 1 - \sqrt{(\beta - 1)^2 + (\alpha - 1)^2 + (r - 1)^2}$ where $\beta = \frac{\bar{y}_s}{\bar{y}_o}, \alpha = \frac{\sigma_s}{\sigma_o}$ and r is the correlation coefficient.

Notes: *The subscripts* s *and* o *of* y *represent simulated and observed data series.*

The PBIAS is a simple yet very useful measure. It indicates the percentage by which the model estimated values are on average larger or smaller than the observed values. In other words, it indicates over or underestimation (in percentage) of the simulated variable (for example river discharge) by the model compared to the observation. Note that for a correct interpretation of over or underestimation, it is necessary to know whether the error is estimated as $(y_s - y_o)$ or $(y_o - y_s)$, where y_o and y_s are the observed and simulated values, respectively. If the former expression is used (as in the equations given in Table 9.1), a positive PBIAS means overestimation and a negative PBIAS means underestimation. However, PBIAS alone is inadequate to comprehend the goodness-of-fit between the two data series, and its major limitation is that the positive and negative errors may cancel each other resulting in a low PBIAS, which can be misleading because individual absolute errors can still be high.

The MAE prevents negative and positive errors to cancel each other, but obviously does not tell the whole story. Another way to prevent the negative and positive errors cancelling effect is using the squared of the errors as in the MSE or RMSE.

The coefficient of correlation (r), which has its root in the works of Francis Galton and Karl Pearson (see, e.g. Ellenberg, 2014; Stewart, 2019), is one of the earliest and still commonly used statistical measures to represent the linear relationship between two data series. Its value ranges between -1 and $+1$ representing a negative or positive relationship. Higher the r value (negative or positive), stronger the relationship, and a value of 0 means that the two data sets are linearly unrelated to each other. Note however that a perfect correlation does not necessarily mean a cause and effect. The coefficient of determination is given by the square of r, so r^2 (also written R^2) and is interpreted as the fraction of the variance in the data described by the model (see, e.g. Moriasi et al., 2007; Spiegel et al., 2013). It does not however represent the bias between the data sets. As a result, the model may systematically over or underestimate the observation data and can still have a good r^2 (Krause et al., 2005). The r^2 together with PBIAS give a much better information about model performance. My favourite example to illustrate this is with a river flow simulation using two different values of roughness (Fig. 9.2). In a river channel the flow velocity is inversely proportional to the channel roughness represented by, for example, Manning's roughness coefficient (see Chapter 7). Keeping other things constant (such as channel cross-section, slope, and inflow discharge), if the Manning's roughness value is increased, the water depth in the channel will increase to convey the same discharge with a reduced velocity. So, if we compare two time series of water levels at a certain section of the river simulated with two different roughness values, the two time series will have a systematic bias, but they can be very well correlated with r^2 close to 1.0 (Fig. 9.2).

The NSE (Nash and Sutcliffe, 1970) uses the ratio of the MSE to the square of the standard deviation (variance) of the observed data to express the model performance. Obviously smaller MSE means a better performance, but how small or large a value is, is a relative thing, and so here the MSE is expressed relative to the observed variance. The NSE values of 1 and 0 give significant meaning. NSE $=1$ (which is the upper limit for NSE) basically means that the two data series are identical, and so the model reproduced the data perfectly. However, NSE $=0$ has a very different meaning than when $r^2=0$. While $r^2=0$ tells that the two data series are linearly unrelated, NSE $=0$ signifies that the model is as good as using the constant mean of the observation data series, because if every $y_{s,i}$ is equal to the mean of the observation, then MSE $=\sigma^2$ (see the equation in Table 9.1). That means, if NSE is less than 0, using the mean of the observation is a better prediction than the model estimates. Note that NSE can be a large negative value; it approaches to $-\infty$ if σ approaches to 0 (division by zero). But is it really so that with NSE <0 the model is less worthy than just using a constant mean value? Not necessarily. This is true only for the 'overall performance over the entire data series', but for parts of the data series the model may still be performing a lot better (so do not throw your model into the dustbin as yet!). For example, the part of the time series where the model is performing well or performing terribly bad could be the dry season or wet season, or beginning of the wet season,

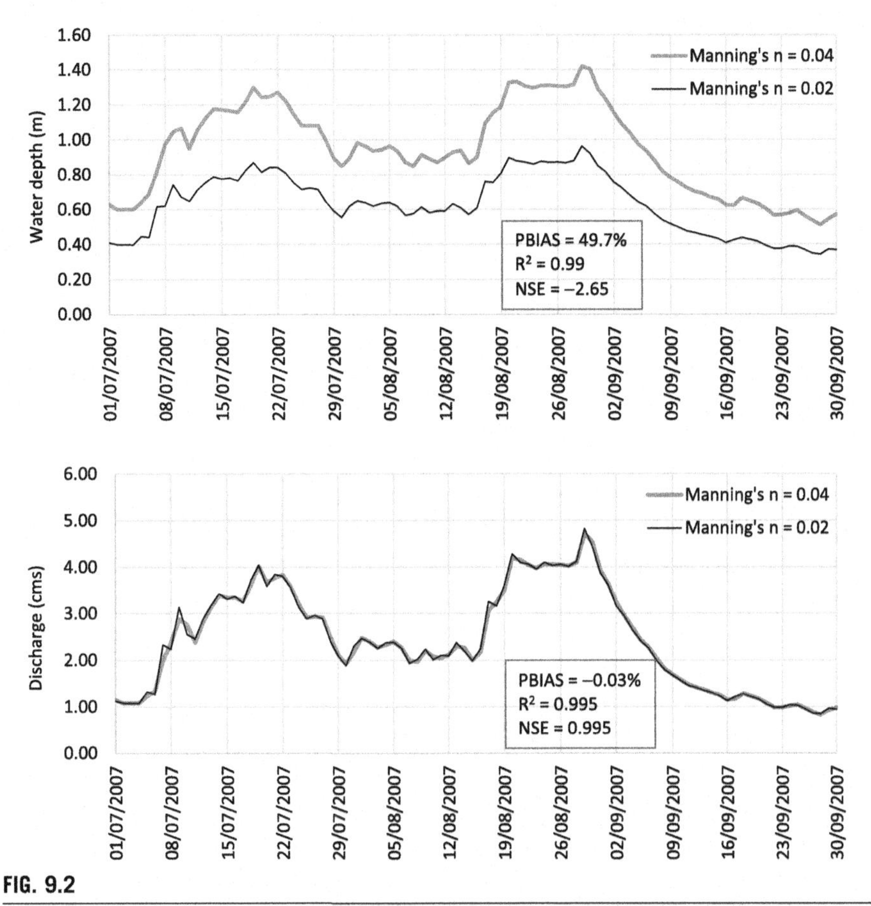

FIG. 9.2

Water level (top) and discharge (bottom) simulation using a hydrodynamic model with two Manning's roughness values ($n=0.04$, and 0.02) with same upstream discharge boundary condition.

or recession part of the wet season, or some particular year. Sometimes a poor model performance could be because of the timing difference (lag) between the model estimated and observed data, and such type of error may have a better possibility to be corrected if properly identified (as illustrated in Fig. 9.2). As mentioned above, $r^2=1$ and $\text{NSE}=1$ also do not mean the same thing. NSE becomes 1 only if two time series are identical, but r^2 can be 1 ever if there is a large bias between the two data sets as long as they are perfectly linearly correlated.

In NSE, the normalization of MSE by the variance of the observed data implies that the simplest imaginable model with the constant mean of the observed data as its estimate (as discussed above) is used as the reference model for normalization. One consequence of this is that NSE may not be a reliable metric to compare model performance across different case studies as argued by Schaefli and Gupta (2007). They also proposed the idea of using a different benchmark (or reference) for NSE estimation for comparing different catchment results. In this sense the NSE may be redefined as

$$\text{NSE} = 1 - \frac{\text{MSE}}{\text{MSE}_{\text{BM}}} \tag{9.1}$$

where MSE_{BM} is the mean squared error of the benchmark model. Such a benchmark model should be appropriately established depending on the prevailing hydrological conditions with respect to the runoff simulation in the catchment.

The use of the squared error which prevents the positive and negative error cancellation also has another consequence, which is that it amplifies the large error values. Such an effect may not be desirable when the model performance with respect to simulating low flows is more important. Because of this, the modified NSE (MNSE) is also used in which the value of the exponent $p = 1$ is commonly used, but $1 \leq p \leq 2$ may also be used.

The decomposition of NSE (Eq. 9.2) shows that it includes three parameters related to the mean bias, standard deviation and correlation, however the decomposition is limited because the standard deviation and correlation terms are not independent (Gupta et al., 2009; Clark et al., 2021).

$$\text{NSE} = -\left(\frac{\mu_s - \mu_o}{\sigma_o}\right)^2 - \left(\frac{\sigma_s}{\sigma_o}\right)^2 + \frac{2\sigma_s r}{\sigma_o} \tag{9.2}$$

where μ_o and μ_s are the mean values and σ_o and σ_s are standard deviations of the observed and simulated data sets, respectively and r is the correlation coefficient.

The KGE proposed by Gupta et al. (2009) also includes the same three parameters as NSE, but derived simply as the Euclidean distance computed using the coordinates of the three parameters (the mean bias, standard deviation and correlation). It has been increasingly used as an alternative to the NSE. More discussion about NSE and KGE metrics can be found in Gupta and Kling (2011), Lamontagne et al. (2020), and Clark et al. (2021).

9.2.3 Temporal scales

Another thing about model performance evaluation is the temporal scale of the data used for computing the metrics. Most hydrological models are run on a daily time step, particularly for a continuous long-term simulation. That is in part because precipitation and temperature, the basic meteorological inputs to hydrological models, are mostly recorded as daily sum (precipitation) and daily average (temperature) values. Same is the case with the observed discharge data for model calibration. Accordingly, the model performance metrics are computed on the daily time series data. In an event-based or short-term simulation, a smaller time step (e.g. hourly or even minutes) are commonly used. In the daily time step simulation, it does not mean that we should evaluate the performance only on daily time series. Analysis of model results on monthly, seasonal and annual time scales also give a lot of useful indications about the model performance. Comparison on the annual scale is the key to understand model performance in terms of the catchment water balance. The Change in storage (= Precipitation – Evaporation – Runoff) may vary by a small percentage (negative or positive) year to year. When averaged over several years, the change in storage is expected to be marginal or close to zero, unless there is a systematic change taking place in the catchment (e.g. groundwater storage) or a large error in the model or data.

In hydrology seasons matter. In many regions, four seasons are well distinguished in which hydrological dynamics also vary. Depending on the rainfall periods, wet or dry seasons are also commonly referred to in hydrology. In catchments with snow precipitation, snow accumulation and snowmelt are important distinctions. In river basins in South East Asia, for example, if the interest is to see how the hydrological model responses to the monsoon rainfall, it may not make sense to split the monsoon period or combine it with other seasons. Similarly, if the interest is also to see how good is the model performance in the dry season, it is better to separate the evaluation for dry and wet seasons. Separation of dry and wet periods for model evaluation is also helpful because the indicators based on mean squared error puts larger weight to the large error terms which are more likely in the wet period simulation. Assessment of the model performance on the monthly time scale usually gives a good indication of seasonal differences in the model response. Note also that for some applications, monthly data may be adequate, for example season drought assessment, basin scale water resources planning, and climate change impacts on future water resources. In this regard, models that may not be good enough on daily time scale but good on monthly time scale are still useful. So for a continuous simulation model, performance evaluations on annual, monthly or seasonal, and daily time scales are all useful.

9.3 Model calibration and validation

As we have seen in Fig. 9.1, a hydrological model consists of several components each representing a certain process or processes (discussed in Chapters 3–8) described by a set of equations. A lot of these equations have parameters whose values may be unknown or cannot be determined by field measurements. Obviously different values of these parameters usually result in different model outputs for the same model input data. Thus, it is a part of the model setup process to select a right set of values for these parameters. In a nutshell, model calibration is about adjusting the values for these unknown parameters in order to obtain a good match between the simulated time series of a target output variable and corresponding values from observation. The outcome of the calibration process is thus a model with a set of parameter values (or sometimes multiple sets of parameter values) that represent the catchment processes in order to appropriately respond to the input data (or forcing).

9.3.1 Calibration procedure

To implement the calibration process, broadly four major steps can be distinguished:

1. Selection of the output variable or variables for which observation data are available for the same period as the input data.
2. Selection of unknown model parameters, also called 'calibration parameters', and their likely value ranges (constraints).
3. Defining a function or functions, called the 'objective function' to evaluate the match or goodness-of-fit between the model simulation and observation of the target variable(s).
4. Executing the procedures from selecting parameter values from the specified parameter value ranges, evaluation of the objective function(s) and finding an optimal set or sets of parameter values based on the evaluated objective function(s).

As discussed in Section 9.2, river discharge is the standard output variable used for catchment model calibration. However, increasingly benefits of other variables, e.g. evapotranspiration, soil moisture and snow cover, along with the discharge in model calibration have been tested (e.g. Finger et al., 2011; Rientjes et al., 2013; Kunnath-Poovakka et al., 2016; Sirisena et al., 2020; Shah et al., 2021). Calibration using two or more variables essentially means multivariable or multiobjective calibration and requires new ways to show added value in calibration and to design criteria for finding the optimal parameter sets (step 4) (see, e.g., Gupta et al., 1998; Rientjes et al., 2013; Finger et al., 2015; Molina-Navarro et al., 2017; Sirisena et al., 2020; Shah et al., 2021).

Success in model calibration depends a lot on the selection of the calibration variables, their value ranges and even the starting values of these variables. Too many parameters and too wide value ranges usually result in less effective optimum selection of parameter values. On the other hand, with too few parameters and too narrow value ranges, the calibration is likely to miss the optimal parameter space. One reason why a distributed model dose not always perform better than its lumped counterpart is in part due to reduced effectiveness of the calibration (optimization) algorithms to search from the large parameter space. See for lumped verses distributed calibration in Kling and Gupta (2009). Often sensitivity analysis of the model parameters to the output variable is necessary to select the parameters more objectively. Sensitivity analysis may be influenced by the initial conditions of the model state variables, such as soil moisture and groundwater storage. It may be necessary to carry out parameter sensitivity using different likely initial conditions or sensitivity of initial conditions (see, e.g., Mazzilli et al., 2012). Modeller's experience and results from similar studies play a vital role in selecting parameters wisely.

The objective function (step 3) is basically the model performance metrics discussed in Section 9.2. We also discussed in Section 9.2 that model performance evaluation using only one metric is often inadequate and sometimes even misleading, so the use of two metrics is more appropriate. This however leads to the case of multiobjective calibration and similar to the multivariable calibration it requires additional criteria for selecting optimal parameter set or sets. More detailed discussion on the multivariable or multiobjective calibration can be found in Gupta et al. (1998) and Vrugt et al. (2006).

Model calibration is essentially an optimization problem and there are host of techniques for efficient sampling of parameter values and evaluation of optimum parameter sets. There are also different calibration tools available, e.g. SWAT-CUP for calibration of SWAT model (Abbaspour, 2015) and the calibration module of HEC-HMS (Feldman, 2000). Some calibration techniques also include model uncertainty analysis. Commonly used such techniques are Generalized Likelihood Uncertainty Estimation (GLUE) (Beven and Binley, 1992; Beven and Freer, 2001), Markov Chain Monte Carlo (MCMC) and Sequential Uncertainty Fitting Algorithm (SUFI2) (Abbaspour et al., 2004). All these four methods are included in the SWAT-CUP. Detailed descriptions of these tools and techniques are not intended in this chapter.

Once the model is satisfactorily calibrated, model validation is usually carried out, which refers to evaluating the model performance with the identified optimal parameter set on a different time period data set not included in the calibration run.

9.3.2 Manual vs automatic calibration

Broadly there are two approaches to perform the calibration procedure: 'manual' or 'automatic'. In automatic calibration, the computer programme performs the whole procedure from selecting parameter values (within the specified parameters and their value ranges), evaluation of model result from each parameter set based on the defined objective function and identification of the best (or most optimal) parameter set, without the user requiring to intervene. In manual calibration, it is the user who decides what parameter values to select for the next run based on the model performance with the current parameter set (user selected) until a satisfactory result is achieved. There are a host of techniques and tools to carry out automatic calibration, but it is from the manual calibration we learn the most about the model and how the model responds to parameters and input data. I always encourage to first attempt manual calibration to diagnose major issues (as pointed out in Section 9.3.3) and only then attempt automatic calibration for more detailed searching and 'fine tuning', so to speak. Use of both manual and automatic calibrations as complementary is the best way to both obtaining the better calibration and better understanding the model (see, e.g. Boyle et al., 2000). In fact, before attempting any calibration, we should first evaluate the model results with the default parameter values (if the model has such). Note that imperfect parameter values are only part of the problem.

But what makes a manual calibration—actually any calibration—a success? On what basis do we select parameters and their values for manual calibration? In a way it is an art as much as science to calibrate a model effectively. First, we need to identify the problem (diagnosis) of course, which are described in Section 9.3.3.

Then if we want to attempt to resolve the problem through model parameters (calibration), the main guiding questions should be the following:

1. What do we expect to happen in the model output and or any other fluxes the model simulates due to the parameter value we are attempting to change?
2. Why should there be the type of change we expect in the model output by changing that parameter value (in either direction, positive or negative)?

For example, for the first question, if we want to change the curve number (CN) value (see Chapter 4), what do we expect to happen in the river discharge as the model output. Do we expect the discharge to increase, decrease, delay in time etc.? We can ask the same question for groundwater recharge or evaporation. Do we expect the recharge to increase or decrease or no effect by increasing the CN value, for example? Similarly, what do we expect to happen in the peak discharge by changing the time of concentration (a parameter of a surface flow routing method), or by changing the groundwater reservoir coefficient, etc.? Then the second question is 'why' is that so? For example, why should the river flow from a rainfall event increase if the CN value is increased? It may turn out, after we perform the model run, that the answers we had in the above two questions were wrong. Which is fine and normal. That is why it is also a learning process, and in the next trial with the same parameter, we may have a different answer. This way we are not only likely to get good calibration quicker but also understand the model better.

9.3.3 'What if' scenarios for catchment model calibration

Catchment model calibration is always a trial and error procedure, and even in automatic calibration, we are unlikely to arrive at satisfactory or best result in one go. Below I have listed a number of problem scenarios that are commonly encountered while calibrating a catchment model and described the likely sources of the problem and possible ways to solve them. These are grouped into analyses on annual time scale, monthly time scale and daily time scale.

On annual time scale

A. The model highly overestimates or underestimates the catchment discharge ($Q_s > Q_o$ or $Q_s < Q_o$) with a large PBIAS (negative or positive), and the biases are more or less consistent year by year.

- There are three most likely reasons for this scenario: (1) Inadequate, erroneous or uncertain precipitation data, (2) Imperfect estimation of evaporation (likely due to erroneous or inadequate weather input, e.g. temperature, or inappropriate interception and evapotranspiration parameters), or (3) Inappropriate groundwater related parameters. There can be always a combination of reasons including the erroneous discharge data used for comparison. However, often there may be one dominant source of uncertainty, which may be more easily detected.
- If the over or underestimation is really high, precipitation may be more likely reason, but not necessarily. The evaporation to precipitation ratio may give

some clue whether the estimated evaporation is within an expected range. But of course to check this, some knowledge or additional information about the hydrology of the catchment is necessary. If the model produces separate components of runoff (surface, lateral and baseflow), the proportions of these components may also be helpful to detect whether the problem is groundwater model related.

- It is logical to check the input data (precipitation and temperature) first in this case. How many precipitation gauges are available? Are they well distributed over the catchment area or are mostly located in one part of the catchment? If the model is a distributed type, how is the rainfall interpolated into the model grid cells? If the model is a semidistributed type, how is the rainfall assigned to each subcatchment? These are the type of questions that help to assess how good is the input data and what can be done to improve the input. Same questions may be asked for other weather data including temperature.
- If the rain gauge density is low or not well distributed over the catchment, interpolated rainfall input usually works better than taking it from the nearest station, and particularly in mountainous catchments elevation based interpolation or the application of lapse rates may be necessary (see e.g. Masih et al., 2011; Sirisena et al., 2018; Nazeer et al., 2022).
- If the precipitation includes snowfall, the over/underestimation can very well be snow related. Inaccurate snow/rain separation or snowmelt computation are likely due to insufficient temperature data, but can also be due to other snowmelt related parameter errors. In a mountainous catchment with steep gradient, large temperature variation exists, which directly impacts snow modelling. So, the application of a temperature lapse rate usually improves the simulation.
- A catchment may lose water to other catchments through deep groundwater recharge if the deep groundwater aquifer is not connected to upstream of the river system where the discharge is measured as the catchment outflow. To take this into account, in some catchment models the lower (deeper) groundwater layer is not connected to the river in the catchment, which means the recharge to this groundwater layer is a loss for the catchment water balance (see, e.g., Feldman, 2000 and Neitsch et al., 2011). So, the parameter related to recharge to the deep groundwater layer may also be related to over/underestimation. In some models, a correction for input precipitation, particularly for snowfall, is also allowed (see, e.g., Bergström, 1992 and Burek et al., 2013). However, such a correction should be applied judiciously.

B. The model overestimates or underestimates the catchment discharge ($Q_s > Q_o$ or $Q_s < Q_o$) with a large PBIAS (negative or positive), but it is caused by high biases in few specific years.

- If the high over/underestimation is not every year more or less consistently but only few specific years, then it is necessary to check if these specific years are also the (1) extreme wet year (or one of the extreme wet years), (2) extreme dry years (or one of the extreme dry years), or just a (3) normal year.

- If it is one of the extreme wet year, the problem is most likely either in the rainfall input or observed discharge or both. High rainfall values that are recorded in some gauges, for example, may be due to a local storm, but if the rainfall from the station is input to a large subcatchment or more than one subcatchment, the rainfall input is overestimated in those catchments and consequently the model overestimates the runoff. So, we should look into the rainfall data in that year. If such an issue is identified, there may be other ways to estimate precipitation for those catchments, e.g. using different interpolation techniques, or using more rainfall stations in the interpolation. Sometimes rainfall may be wrongly recorded, e.g. wrongly placed decimal point, etc.
- The problem in an extreme wet year can well be discharge measurement too. As said earlier, daily discharge data are commonly estimated using a rating curve. Rating curve accuracy is likely low for high discharge values as those values are often an extrapolation in the rating curve. However, such a high flood discharge if occures only once or twice in the year, for example, is less likely to affect significantly on the annual scale. This effect is more important in the assessment on the daily time step.
- If the problem is in the extreme dry year, rainfall input is again the main suspect. Low rain gauge density or unevenly distributed rain gauges may impact on both high and low rainfall periods. The effect of discharge measurement data may be less of a reason in this case, but of course it cannot be entirely excluded.
- Model parameters are also likely to play substantial role in this case. If the parameter values are biased towards high rainfall periods/years, it may affect adversely in the low rainfall periods/years. One simple example is with the SCS CN method. In a high rainfall year, the model likely needs relatively smaller CN number than in a low rainfall year. So, if the parameter is biased in favour of the high rainfall year, the runoff in the low rainfall year may be reduced more than it is necessary.

C. Large variation in annual biases (discharge) across subcatchments

- This is a commonly encountered issue in catchment modelling if the catchment has several or many subcatchments, and particularly if the input rainfall gauges are unevenly distributed in the catchment. It will be useful to find out whether the catchments with poorer model performance are also the catchments without rain gauges and temperature measurements. So, the first thing is to try to improve the input data same as discussed in scenarios A and B above by interpolation, lapse rate, etc.
- Sometimes it may also be that there are small reservoirs or water diversions that are not included in the model. Such small reservoirs or diversions may not have notable effects on large catchments but can be substantial for small catchments.
- The calibration process is also important here. If there are multiple cascading catchments (one or more subcatchments draining to a catchment

downstream), and if a simultaneous calibration of the entire catchment is not helpful, it is better to first calibrate the catchments which do not have a upstream catchment. Then keep the calibrated parameters of those catchments while calibrating the downstream catchments.

On monthly or seasonal time scale

D. The model highly overestimates or underestimates the catchment discharge ($Q_s > Q_o$ or $Q_s < Q_o$) with a large PBIAS (negative or positive), and the biases are more or less consistent throughout the year.

- This scenario is to a large extent similar to scenario A. If the analysis on the annual scale is carried out and solution is found (partially or fully), that should be reflected here too.

E. The model highly overestimates or underestimates the catchment discharge ($Q_s > Q_o$ or $Q_s < Q_o$) with a large PBIAS (negative or positive), but the biases are mainly in the high flow months.

- This is also related to A above, but shows that the biases are not uniform within the year. The possible improvements in input data identified in A may also solve this fully or partly, but if it still remains, the problem may be on model parameters or calibration data as well.
- Parameters related to evaporation and unsaturated zone (particularly separation of surface flow volume and soil infiltration) are the areas to look for. For example, overestimation of flow on high rainfall months are likely due to a relatively large proportion of surface flow than infiltration (e.g. a large CN value or low infiltration capacity). On the other hand, underestimation of high flow discharge is likely due to too low surface flow and too high infiltration resulting in high evapotranspiration and groundwater recharge in conjunction with faulty groundwater parameters. Note that related to evapotranspiration and infiltration (or CN) also means land cover and soil data.
- Under or overestimation of high flow in a long-term continuous simulation is a commonly encountered issue. The problem is that trying to solve this through parameters is likely to adversely affect the simulation in the low flow period. Thus, in the end it comes down to finding a balance between model performance under various conditions.

F. The model highly overestimates or underestimates the catchment discharge ($Q_s > Q_o$ or $Q_s < Q_o$) with a large PBIAS (negative or positive), but the biases are mainly in low flow months.

- Apart from the reasons related to input data discussed in scenario A, evaporation and groundwater related parameters are most likely reasons for this scenario. Unsaturated zone parameters also affect here because they affect evapotranspiration and groundwater recharge.
- During low flow periods almost all the river flow comes from groundwater, except in catchments with snow or glaciers or if the river is fed by lakes or reservoirs. Thus, groundwater recharge during the wet period and

groundwater response parameters play a major role for the low flow discharge. A fast groundwater response empties groundwater storage quickly and as a result baseflow cannot sustain over a long dry period, and a too slow groundwater response may result in a low baseflow contribution during the wet period. Finding this balance can be difficult if there is only one groundwater layer. If the model has two or more groundwater layers, varying the response time parameters in the two layers generally helps (typically a faster response from the upper/shallow groundwater layer and slower response from the lower/deep layer).

- Too high or too low evapotranspiration, both in wet and dry seasons, is likely to be related to inaccurate land cover and soil data/parameters apart from the weather data.

G. There is a shift in the peak discharge month between the simulated and observed data

- In a snow/ice free catchment, shift in the peak discharge month between observed and simulated data is not usually expected unless multiple months have discharge of similar magnitudes. But if it exists it may suggest a substantial mismatch between the rainfall input and discharge data used for comparison. This effect may be seen in scenario E as well. First things to check are the input data (particularly rainfall) and discharge data for those months.
- If there are storage reservoir(s) in the catchment that are not included in the model, such a scenario is not a surprise. Reservoirs are mostly used to store water in the wet period and supply more water during the dry period. So, we should check for any existence of reservoirs or lakes in the catchment and whether they needs to be represented in the model.
- In a catchment with snow/ice, it is very likely related to temperature and other weather data and snow related parameters. Modelling studies in the snow/glacier rich river basins like Kabul and Upper Indus (Nasery, 2017; Nazeer et al., 2022) show that without reasonable temperature data and snow/ice melt model, it is unlikely to obtain flow simulation anywhere close to the observed flow.

H. Large variation in model performance across subcatchments

- Same as in scenario C. With the analysis on monthly basis, it is usually easier to detect the main responsible factors and potential solutions.

On daily time scale

I. The model highly overestimates or underestimates the peak discharges

- This is a common issue often encountered in catchment modelling. Models may capture some peak flows well but may fail to capture others (as also seen in Fig. 9.3). It is likely to be caused by inadequate and erroneous data as well as limitations of the model structure and inappropriate model parameters. Discharge data (used for comparison or calibration) is likely to be more uncertain at high flows (because of extrapolation in the rating curve).

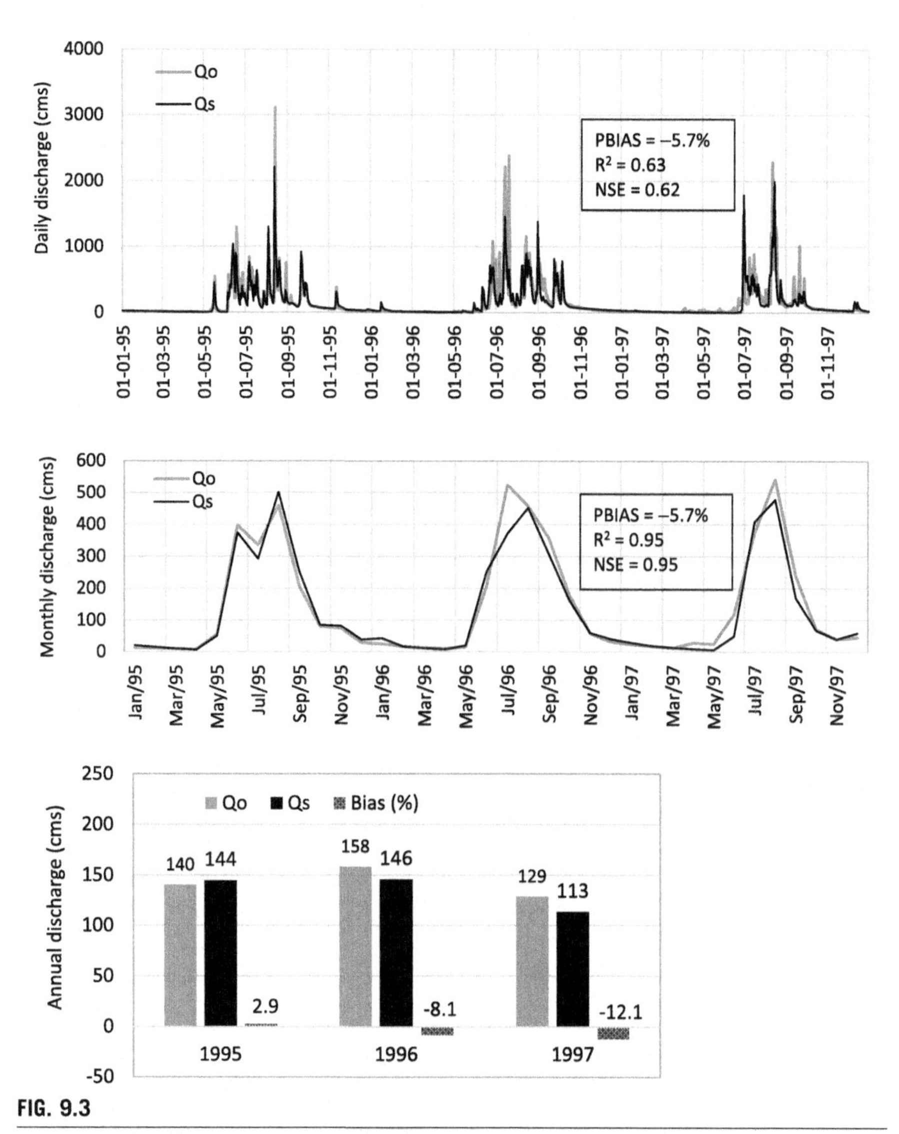

FIG. 9.3

An example showing the comparison of observed and model simulated discharge on daily, monthly and annual time scales. The PBIAS is calculated using the equation presented in Table 9.1. Accordingly, positive and negative PBIAS values indicate an overestimation and underestimation, respectively, of the discharge by the model.

Moreover, the water levels are usually measured twice a day, and as the flood peak discharge may not last for the whole day, the actual peak may be missed, or if the peak is captured, the daily average flow may be overestimated.
- Model spatial resolution (grid size or subcatchment size) is also important to get the peak flow right. A coarse spatial resolution model usually results in reduced peak discharge as it averages over a large area, but it is not always the case because it also depends on how the rainfall is estimated for the area.
- Model parameters mostly responsible are related to surface runoff and soil infiltration (unsaturated zone). Accordingly, land use and soil data also play a significant role. A small increase in the impervious area may result in significant increase in the surface flow volume and peak flow, for example.
- As discussed in scenario E, trying to improve the model performance on peak flow simulation through parameter adjustment may result in decreased model performance on low flows.

J. The model highly overestimates or underestimates the low flow discharges
- Same as in scenario F.

K. Peak flow discharge values simulated and observed are not on the same day (timing error of peak discharge)
- The timing error of peak flow on a daily or subdaily time step is an often encountered issue, which can also be observed in Fig. 9.3. If there is no mismatch between the model input data and observed discharge used for calibration or comparison, the timing error issue is generally possible to solve to some extent through model parameters and spatial resolution.
- Routing parameters (surface flow routing and river routing) are likely to be responsible to some extent and relatively easy ones to adjust. For example, depending on the methods used, parameters such as the time of concentration, reservoir routing constant and Manning's roughness coefficient affect the timing aspect of the simulated hydrograph. However, the effect of the routing parameters is less effective on a daily and larger time step.
- The unsaturated zone parameters for the partitioning of rainfall into surface flow volume and infiltration can also play a major role particularly through the effect of antecedent soil moisture conditions.
- Model spatial resolution is also important here in the similar way as discussed in scenario I.

L. There are large biases in the rising and receding phases of the wet-period.
- This is also a commonly encountered problem in catchment model simulation. Apart from the input data (mainly rainfall), unsaturated zone parameters related to antecedent soil moisture may be responsible for the mismatch in the rising phase. Accurate accounting of the antecedent moisture condition is difficult because it is influenced by parameters that are usually less known and spatially heterogenous, such as soil texture, hydraulic conductivity, depth etc.
- The problem may also be in the observed discharge used in calibration or comparison. Note that water depth to discharge rating curves are hysteresis, with different relationships for the rising and receding floods. Thus, if the

rating curve is not corrected for hysteresis effect, discharge error is inevitable during flood rising/receding phases. However, such an effect is more prominent on a subdaily time step than daily time step simulation.
- In a catchment with snow precipitation, it may very well be related to inadequate temperature data (and other weather data) and snow accumulation and melt parameters. See more on this in scenarios M and N.

M. The model overestimates or underestimates the snowmelt period discharge (if the catchment has snow precipitation).
- This is likely related to inadequate temperature data (and other weather data) as well as snow accumulation and melt parameters. Particularly in a catchment with steep slope, temperature varies substantially within a small horizontal distance. In most cases temperature lapse rate is inevitable to be applied, which usually improves the simulation.
- It is a standard practice to use a constant temperature threshold to distinguish a given day's precipitation as snow or rain. In reality, the threshold may vary and an inappropriate threshold can introduce significant errors in snow accumulation and runoff. Same can be said about the snowmelt threshold temperature and melt rate factor (in the case of the temperature index method).
- Another major issue when snow modelling is involved is the uneven distribution of snow within a catchment or grid cell. If the precipitation is snow, it may not have a uniform depth over the entire subcatchment, and so the amount of snowmelt produced at different parts of the same subcatchment or grid cell may be different. So it may be helpful to check how this issue is treated (or not) in the model and if the related parameter(s) may have affected in the over/underestimation of the runoff.

N. The snowmelt discharge starts too early or too late
- Inadequate temperature and other weather data and the temperature threshold for snowmelt (if the snowmelt is based on the temperature index method) are the most likely reasons for this scenario.
- Other factors discussed in scenario M may also have effects to some extent.

O. Large variation in model performance across subcatchments (usually better in larger subcatchments than smaller ones)
- Same as discussed in scenario C.

9.4 How uncertain is my model result?

While the use of performance metrics indicates how good the model has simulated the output variable of interest from the known set of input data and best knowledge of the model parameter values (that are optimally selected in calibration), by uncertainty assessment the modeller attempts to indicate their level of confidence in the model for its use with input data and situations beyond calibration and validation. It is to acknowledge that model results are always subject to some uncertainty, because (1) the model is always a simplification of the real system (that is the catchment in our case), (2) there always remain some parameters whose values are not well known, (3) data are mostly inadequate to represent spatial and temporal variability at intended details, and some inaccuracy in the measurements cannot be ruled out, and (4) even if the model performed well in calibration and validation, it cannot be guaranteed that the model works equally well with every new input data set.

Uncertainty in modelling (theory and applications) is a huge field of research and there is plenty of published literature in the subject. Detailed introduction of different methods and tools for uncertainty assessment is not intended in this book, for which some of the cited references may be consulted. In the following two sections I attempt to give a very brief overview of concepts essential for understanding uncertainty and approaches to uncertainty assessment in catchment modelling.

9.4.1 Understanding uncertainty in catchment modelling

While the importance of knowing uncertainty in model prediction is well recognized, uncertainty assessment is not always a part of hydrological modelling studies. This is in part because in the catchment modelling context it is not easy to understand uncertainty, to quantify it sensibly and to interpret it meaningfully. So before trying to explain what and how we may assess uncertainty in catchment modelling, let us look at few simple examples.

The first example is about measurement uncertainty. Suppose we want to measure the discharge in a river. There are different ways the discharge may be measured these days, but the basic principle is that discharge is the velocity times the area of the river channel cross-section. So, if we can measure the velocity of the flow and area of the cross-section occupied by the flow at given time, we can compute the discharge. But we know that velocity is not constant throughout the cross-section. It is different at different heights from the channel bottom, for example, and also varies across the channel width depending on the cross-section shape and channel bed materials (e.g. different on thick vegetated surface than on find sand, for example). So, the standard practice is to divide the channel into a number of vertical sections and measure velocities at different depths in each section. Then with the area of each section and averaged velocity, estimate the discharge through the section and aggregate the discharge values from all the sections to estimate the total discharge through the channel. If we repeat this process over and over again, we can be pretty sure that there will be some differences in our estimates each time even if it is carried out by the same people. If we repeat it say 100 times, we get 100 estimates (does not mean that all 100 will be different) of the same discharge, from which we calculate the standard deviation (also called the standard error), observe the distribution, etc. This allows us to estimate the popularly used 'confidence interval' and 'confidence level' in our estimate, and this is one way of expressing measurement uncertainty (a detailed procedure can be found in most standard statistics text books).

The second example is about guessing the temperature of tomorrow. Suppose we are asked to guess the temperature of tomorrow at 8 am. Of course, we cannot know it for sure, so we can give our prediction not as one number but as a range (or interval). To do this a lot of thinking goes in our mental model, and we probably question: What was the temperature in the morning today? What was it yesterday? Was the temperature of last several days relatively constant or was it deviating day by day? Et cetera. Basing on the same information and data and the mental model for processing, one thing is sure about our prediction interval. If we use a large interval, for example knowing today's temperature was 10 °C and we say tomorrow will be between 8 °C and 12 °C, we will have a higher confidence in our prediction interval than if we use a smaller interval, say between 9.5 °C and 10.5 °C. This simple example reminds us that the confidence interval and confidence levels are actually relative thing. If we choose a large interval to express our uncertainty we will have higher confidence that our prediction will fall within the predicted interval, and conversely, if we choose a smaller interval we will have lower confidence. However, the interval for the same confidence level also vary if we use different sources of data and inference model.

Estimating uncertainty in catchment modelling is in many ways similar to the above simple examples, but the major difference is that it is a lot more complicated process. As we already know a catchment model consists of a number of component models (Fig. 9.1), all of them can be of different levels of details, and require different types of data for representing the catchment and the weather. To give one example of the catchment model, let us consider the unsaturated zone model based on the SCS CN method (see Chapter 4). We know that although CN values are estimated based on land cover and soil data, its estimate is always an approximation. One would estimate a different value if the land cover and soil data used are from different sources or have different spatial resolution, which essentially means there is some uncertainly in the CN value we assign to any catchment. Say we express the uncertainty in the CN value using an interval, e.g. between 60 and 80 instead of a single value 70. Keeping every other thing the same, if we run the model with CN 60, 70, and 80 we will get three different sets of runoff simulation. We can also rerun with any other CN values between 60 and 80. These model runs give us not just one value (per time step) of runoff but a range of values, which represent the uncertainty in the simulated runoff due to the uncertain parameter CN. This is just with one parameter of one model component. There are many model components and uncertain parameters in a hydrological model, but the key concept is the same.

There are more sources than just the uncertain parameters that induce uncertainty in catchment modelling, which can be classified as due to (1) model uncertainty including model structure (e.g. lumped, semidistributed or distributed), (2) input data and calibration data uncertainty including spatial and temporal resolution of the data, (3) model parameter uncertainty, and (4) model initial and boundary condition uncertainty.

The model uncertainty is due to the assumptions (and simplifications) in the model equations, model construction (e.g. how different model components are integrated), and other forms of incompleteness in representing the real physical system. Assessment of model structure uncertainty is not commonly done, but the importance of knowing the limitations of the model with respect to what we want to use if for cannot be neglected. Model uncertainty can be reduced by carefully selecting the model depending on the important processes in the catchment. For example, interception method is mostly a lot simplified or lumped in hydrological models. If the catchment we want to model is densely forested, interception can be a significant component and it may require a detailed representation. Model structure also influences on how the input data uncertainty propagates in the model (Montanari and Di Baldassarre, 2013). One way to understand and handle model uncertainty is using ensemble of models, which is particularly useful in forecasting. More about model ensemble techniques can be found in, e.g. Vrugt and Robinson (2007), Diks and Vrugt (2010) and Tyralla and Schumann (2016).

Input data uncertainty can be both from the weather data, e.g. precipitation and temperature, and the data that represent the catchment physical features, e.g. topography and land cover. Weather input data are often scarcely available and can be a significant part of uncertainty (see, e.g., Maskey et al., 2004; McMillan et al., 2011). In principle, input data uncertainty can be included in the uncertainty framework, but is often neglected or considered rather crudely due to obvious complexities as well as lack of sufficient information. Challenges of incorporating model and input uncertainty are discussed in Renard et al. (2010).

The model parameter uncertainty is probably the most commonly reported source of uncertainty in catchment modelling literature. The CN uncertainty discussed above is an example of model parameter uncertainty. The parameter uncertainty and its handling in uncertainty assessment are further discussed in the next section.

The most common initial conditions in a catchment model are the soil moisture and groundwater storage, whose precise values are unknown. A note of caution: the classification may vary in different literature and some sources of uncertainty can be put in one or the other category, but the most important is to understand what they mean and how they can be treated. Initial snow storage, lakes and river water levels can also be considered as types of initial conditions. Typically, models are run with the 'warmup period' (usually one to three years) before the calibration period to reduce the effect of initial condition uncertainty. In a physically based groundwater model, uncertain boundary conditions can have considerable impact on the model result.

9.4.2 How to approach uncertainty assessment in catchment modelling

The example of CN parameter uncertainty discussed in Section 9.4.1 also tells that if we know the uncertainty in model parameters in terms of the 'uncertainty range' or interval, uncertainty in the model results (called 'predictive uncertainty') due to the parameter uncertainty can be estimated by running the model repeatedly with the values of the unknown parameter within its uncertainty range. This procedure is commonly referred to as the Monte Carlo simulation, which is the widely used technique to 'propagate' uncertainty in the input (data and parameter) through the model to estimate uncertainty in the output. In a simple interval representation it does not specify if the parameter values are more likely at certain part of the interval than the rest. In other words, it assumes the values are equally likely within the range. Uncertainty is more commonly represented probabilistically in which probability density function is used to express the difference in likelihood of the parameter within the uncertain range. Accordingly, the uncertainty in the model output can also be represented as a probability density. Note that uncertainty can also be represented using fuzzy membership function or more generally possibility density function and the output uncertainty can be estimated using fuzzy extension principle (e.g. Maskey, 2004; Maskey et al., 2004).

These uncertainty estimation techniques (Monte Carlo simulation and fuzzy extension principle) can be carried out independent of model calibration. The major problem however is that it is not easy or we lack information or criteria to specify realistic uncertainty bound in each uncertainty parameters. As mentioned in Section 9.3.1, model calibration and uncertainty can be combined using the techniques such as GLUE, MCMC and SUFI2. Depending on the availability of data for model calibration and validation, three scenarios can be distinguished to select the type approach to uncertainty assessment in catchment modelling. These are A. When observation data are available for model calibration and validation, B. When observation data are available for calibration but not for validation, and C. When no observation data are available for calibration and validation. How to approach the model calibration and uncertainty assessment in each of these scenarios are described in Table 9.2.

Table 9.2 Approaches to model calibration and uncertainty analysis based on the scenarios of observation data availability.

Scenarios	Approach to model calibration/validation and uncertainty assessment
A. Observation data are available for model calibration and validation.	○ Model calibration and validation with two separate data periods are a commonly used practice in catchment modelling studies. If uncertainty assessment is not necessary, the model may be calibrated as an optimization problem to find an optimal parameter set, and run the simulation with the optimal parameter set on the validation data set. As discussed in Section 9.2.2, selection of the performance evaluation metrics is important and the use of two or more metrics is recommended. ○ Selection of parameters for calibration is best done with sensitivity analysis and modeller's experience. ○ If the result in validation is not as satisfactory as in the calibration, it may be that different hydrological conditions (e.g. relatively wet year, dry year and normal year) are not well represented in the calibration. Reselecting the calibration period may be tried. Another option is to calibration not only for a 'best parameter set' (most optimal), but for several 'good parameter sets' (also called a 'pareto optimal' parameter sets). It is more likely that the one or more of the 'good parameter sets' may work better in the validation set than the one best parameter set.

Continued

Table 9.2 Approaches to model calibration and uncertainty analysis based on the scenarios of observation data availability—*cont'd*

Scenarios	Approach to model calibration/validation and uncertainty assessment
	○ If the uncertainty assessment is required or intended, 'combined calibration and validation' can be carried out using the methods discussed earlier, such as MCMC, GLUE, SUFI2.
B. Observation data are available for model calibration but not for validation.	○ Probably the best option here is to use the combined 'calibration and uncertainty analysis' methods (e.g. MCMC, GLUE, SOFI2). ○ However, if the validation data are not available because the model is intended to use for climate change impact assessment, then the uncertainty assessment conditioned on the present or historical data is likely considered inadequate or unrealistic. In such a case a scenario-based approach or ensemble approach (discussed in C below) is probably a better approach. Note also that in the climate change impact assessment, uncertainty due to uncertain input data is likely to be more dominant than the parameter uncertainty. But the combination of the two may also further exacerbate the prediction uncertainty.
C. Observation data are not available for model calibration and validation.	○ This is not a desirable situation of course to have no data for model calibration and validation. So, we need to exhaust different possible options in order to establish confidence with the model. ○ Luckily uncertainty assessment using the forward modelling approach (such as the Monte Carso simulation or fuzzy extension principle) can still be applied and it is certainly a good way forward. ○ Probably the most important part of the work here is to collect as much information as possible about the parameters of the model and their likely ranges from the literature (preferably from hydrologically comparable catchments) or from any other sources, including expert evaluation. ○ Sensitivity analysis is also important here to select the parameters effectively. ○ The scenarios or the ensemble approach can be very helpful in this case. Scenarios can be developed based on the assessment of likely catchment conditions in which the model may be applied, for example different land use scenarios. Using ensembles of different input data sets is a commonly used technique for climate change impact assessment, in which case the inputs are selected from different global climate model projections. Global precipitation data from various sources are available nowadays. Ensembles can also be developed using these precipitation data sets (see e.g. Sirisena et al., 2018).

References

Abbaspour, K.C., 2015. SWAT-CUP: SWAT Calibration and Uncertainty Programs—A User Manual. Swiss Federal Institute of Aquatic Science and Technology, Duebendorf, Switzerland, Eawag.

Abbaspour, K.C., Johnson, C.A., Van Genuchten, M.T., 2004. Estimating uncertain flow and transport parameters using a sequential uncertainty fitting procedure. Vadose Zone J. 3, 1340–1352.

Bergström, S., 1992. The HBV model—its structure and applications. SMHI RH No 4. Norrköping.

Beven, K.J., Binley, A., 1992. The future of distributed models: model calibration and uncertainty prediction. Hydrol. Process. 6, 279–298.

Beven, K., Freer, J., 2001. Equifinality, data assimilation, and uncertainty estimation in mechanistic modelling of complex environmental systems using the GLUE methodology. J. Hydrol. 249, 11–29.

Boyle, D.P., Gupta, H.V., Sorooshian, S., 2000. Toward improved calibration of hydrologic models: combining the strengths of manual and automatic methods. Water Resour. Res. 36 (12), 3663–3674.

Burek, P., Van der Knijff, J., De Roo, A., 2013. LISFLOOD distributed water balance and flood simulation model, revised user manual. JRC Technical Report 'EUR 26162 EN'.

Chadalawada, J., Babovic, V., 2019. Review and comparison of performance indices for automatic model induction. J. Hydroinf. 21 (1), 13–31. https://doi.org/10.2166/hydro.2017.078.

Clark, M.P., Vogel, R.M., Lamontagne, J.R., Mizukami, N., Knoben, W.J.M., Tang, G., et al., 2021. The abuse of popular performance metrics in hydrologic modeling. Water Resour. Res. 57. https://doi.org/10.1029/2020WR029001, e2020WR029001.

Diks, C.G.H., Vrugt, J.A., 2010. Comparison of point forecast accuracy of model averaging methods in hydrologic applications. Stoch. Environ. Res. Risk Assess. 24, 809–820. https://doi.org/10.1007/s00477-010-0378-z.

Ellenberg, J., 2014. How Not to Be Wrong: The Hidden Maths of Everyday Life. Penguin Books.

Feldman, A.D. (Ed.), 2000. Hydrologic Modelling System HEC-HMS Technical Reference Manual. US Army Corps of Engineers, Hydrologic Engineering Centre, Washington, DC.

Finger, D., Pellicciotti, F., Konz, M., Rimkus, S., Burlando, P., 2011. The value of glacier mass balance, satellite snow cover images, and hourly discharge for improving the performance of a physically based distributed hydrological model. Water Resour. Res. 47, 1–14.

Finger, D., Vis, M., Huss, M., Seibert, J., 2015. The value of multiple data set calibration versus model complexity for improving the performance of hydrological models in mountain catchments. Water Resour. Res. 51, 1939–1958.

Gupta, H.V., Kling, H., 2011. On typical range, sensitivity, and normalization of mean squared error and Nash-Sutcliffe efficiency type metrics. Water Resour. Res. 47, W10601. https://doi.org/10.1029/2011WR010962.

Gupta, H.V., Sorooshian, S., Yapo, P.O., 1998. Toward improved calibration of hydrologic models: multiple and noncommensurable measures of information. Water Resour. Res. 34 (4), 751–763.

Gupta, H.V., Kling, H., Yilmaz, K., Martinez, G., 2009. Decomposition of the mean squared error and NSE performance criteria: implications for improving hydrological modeling. J. Hydrol. 377 (1–2), 80–91.

Hall, M.J., 2001. How well does my model fit the data? J. Hydroinf. 03 (1), 49–55.
Hall, D.K., Riggs, G.A., Salomonson, V.V., DiGirolamo, N.E., Bayr, K.J., 2002. MODIS snow-cover products. Remote Sens. Environ. 83, 181–194.
Kling, H., Gupta, H., 2009. On the development of regionalization relationships for lumped watershed models: the impact of ignoring sub-basin scale variability. J. Hydrol. 373 (3–4), 337–351.
Krause, P., Boyle, D.P., Bäse, F., 2005. Comparison of different efficiency criteria for hydrological model assessment. Adv. Geosci. 5, 89–97. https://doi.org/10.5194/adgeo-5-89-2005.
Kunnath-Poovakka, A., Ryu, D., Renzullo, L.J., George, B., 2016. The efficacy of calibrating hydrologic model using remotely sensed evapotranspiration and soil moisture for streamflow prediction. J. Hydrol. 535, 509–524.
Lamontagne, J.R., Barber, C.A., Vogel, R.M., 2020. Improved estimators of model performance efficiency for skewed hydrologic data. Water Resour. Res. 56. https://doi.org/10.1029/2020WR027101, e2020WR027101.
Masih, I., Maskey, S., Uhlenbrook, S., Smakhtin, V., 2011. Assessing the impact of areal precipitation input on streamflow simulations using the SWAT model. J. Am. Wat. Res. Assoc. (JWARA) 47 (1), 179–195.
Maskey, S., 2004. Modelling Uncertainty in Flood Forecasting Systems. PhD Thesis (ISBN 90 5809 694 7), A.A. Balkema Publishers, Taylor & Francis Group plc, London, p. 178.
Maskey, S., 2019. How can flood modelling advance in the "big data" age? J. Flood Risk Manage. 12. https://doi.org/10.1111/jfr3.12560, e12560.
Maskey, S., Guinot, V., Price, R.K., 2004. Treatment of precipitation uncertainty in rainfall-runoff modelling: a fuzzy set approach. Adv. Water Resour. 27 (9), 889–898.
Maskey, S., Uhlenbrook, S., Ojha, S., 2011. An analysis of snow cover changes in the Himalayan region using MODIS snow products and in-situ temperature data. Clim. Change 108, 391–400.
Mazzilli, N., Guinot, V., Jourde, H., 2012. Sensitivity analysis of conceptual model calibration to initialization bias. Application to karst spring discharge models. Adv. Water Resour. 42, 1–16.
McMillan, H., Jackson, B., Clark, M., Kavetski, D., Woods, R., 2011. Rainfall uncertainty in hydrological modelling: An evaluation of multiplicative error models. J. Hydrol. 400, 83–94. https://doi.org/10.1016/j.jhydrol.2011.01.026.
Miralles, D.G., Holmes, T.R.H., De Jeu, R.A.M., Gash, J.H., Meesters, A.G.C.A., Dolman, A.J., 2011. Global land-surface evaporation estimated from satellite-based observations. Hydrol. Earth Syst. Sci. 15, 453–469. https://doi.org/10.5194/hess-15-453-2011.
Molina-Navarro, E., Andersen, H.E., Nielsen, A., Thodsen, H., Trolle, D., 2017. The impact of the objective function in multi-site and multi-variable calibration of the SWAT model. Environ. Model. Software 93, 255–267.
Montanari, A., Di Baldassarre, G., 2013. Data errors and hydrological modelling: The role of model structure to propagate observation uncertainty. Adv. Water Resour. 51, 498–504.
Moriasi, D.N., Arnold, J.G., Van Liew, M.W., Binger, R.L., Harmel, R.D., Veith, T.L., 2007. Model evaluation guidelines for systematic quantification of accuracy in watershed simulations. Am. Soc. Agric. Biol. Eng. 50, 885–900. https://doi.org/10.13031/2013.23153.
Mu, Q., Zhao, M., Running, S.W., 2011. Improvements to a MODIS global terrestrial evapotranspiration algorithm. Remote Sens. Environ. 115, 1781–1800.
Nasery, M.T., 2017. Estimation of Flow Duration Curves for use in Hydropower Development for ungauged or partially gauged subbasins in the Kabul River basin, Afghanistan. MSc

thesis report (WSE-HWR 17.13), IHE Delft Institute for Water Education, Delft, the Netherlands.

Nash, J.E., Sutcliffe, J.V., 1970. River flow forecasting through conceptual models. J. Hydrol. 10, 282–290.

Nazeer, A., Maskey, S., Skaugen, T., McClain, M.E., 2022. Simulating the hydrological regime of the snow fed and glaciarised Gilgit Basin in the Upper Indus using global precipitation products and a data parsimonious precipitation-runoff model. Sci. Total Environ. 802. https://doi.org/10.1016/j.scitotenv.2021.149872.

Neitsch, S.L., Arnold, J.G., Kiniry, J.R., Williams, J.R. (2011). Soil and Water Assessment Tool Theoretical Documentation Version 2009. Texas Water Resources Institute. Available electronically from https://hdl.handle.net/1969.1/128050.

Notarnicola, C., 2020. Hotspots of snow cover changes in global mountain regions over 2000–2018. Remote Sens. Environ. 2020 (243). https://doi.org/10.1016/j.rse.2020.111781, 111781.

Renard, B., Kavetski, D., Kuczera, G., Thyer, M., Franks, S.W., 2010. Understanding predictive uncertainty in hydrologic modeling: The challenge of identifying input and structural errors. Water Resour. Res. 46, W05521. https://doi.org/10.1029/2009WR008328.

Rientjes, T.H.M., Muthuwatta, L.P., Bos, M.G., Booij, M.J., Bhatti, H.A., 2013. Multi-variable calibration of a semi-distributed hydrological model using streamflow data and satellite-based evapotranspiration. J. Hydrol. 505, 276–290.

Schaefli, B., Gupta, H.V., 2007. Do Nash values have value? Hydrol. Process. 21, 2075–2080.

Shah, S., Duan, Z., Song, X., Li, R., Mao, R., Liu, J., Ma, T., Wang, M., 2021. Evaluating the added value of multi-variable calibration of SWAT with remotely sensed evapotranspiration data for improving hydrological modelling. J. Hydrol. 603 (Part C), 127046. https://doi.org/10.1016/j.jhydrol.2021.127046.

Sirisena, T.A.J.G., Maskey, S., Ranasinghe, R., Babel, M.S., 2018. Effects of different precipitation inputs on streamflow simulation in the Irrawaddy River Basin, Myanmar. J. Hydrol.: Region. Stud. https://doi.org/10.1016/j.ejrh.2018.10.005.

Sirisena, T.A.J.G., Maskey, S., Ranasinghe, R., 2020. Hydrological model calibration with streamflow and remote sensing based evapotranspiration data in a data poor basin. Remote Sens. (Basel) 12 (22), 3768. https://doi.org/10.3390/rs12223768.

Spiegel, M.R., Schiller, J., Srinivasan, R.A., 2013. Schaum's Outline of Probability and Statistics. McGraw Hill.

Stewart, I., 2019. Do Dice Play God? The Mathematics of Uncertainty. Profile Books, London.

Trambauer, P., Dutra, E., Maskey, S., Werner, M., Pappenberger, F., van Beek, L.P.H., Uhlenbrook, S., 2014. Comparison of different evaporation estimates over the African continent. Hydrol. Earth Syst. Sci. 18, 193–212. https://doi.org/10.5194/hess-18-193-2014.

Tyralla, C., Schumann, A.H., 2016. Incorporating structural uncertainty of hydrological models in likelihood functions via an ensemble range approach. Hydrol. Sci. J. 61 (9), 1679–1690. https://doi.org/10.1080/02626667.2016.1164314.

Vrugt, J.A., Robinson, B.A., 2007. Treatment of uncertainty using ensemble methods: comparison of sequential data assimilation and Bayesian model averaging. Water Resour. Res. 43, 1–15.

Vrugt, J.A., Clark, M.P., Diks, C.G.H., Duan, Q., Robinson, B.A., 2006. Multi-objective calibration of forecast ensembles using Bayesianmodel averaging. Geophys. Res. Lett. 33, L19817. https://doi.org/10.1029/2006GL027126.

Index

Note: Page numbers followed by *f* indicate figures, *t* indicate tables and *b* indicate boxes.